Anonymous

The American Rose Culturist

Anonymous

The American Rose Culturist

ISBN/EAN: 9783744690409

Printed in Europe, USA, Canada, Australia, Japan

Cover: Foto ©berggeist007 / pixelio.de

More available books at **www.hansebooks.com**

SAXTON'S COTTAGE AND FARM LIBRARY.

THE

AMERICAN ROSE CULTURIST;

BEING A

PRACTICAL TREATISE

ON THE

PROPAGATION, CULTIVATION, AND MANAGEMENT

OF

THE ROSE

IN ALL SEASONS; WITH A LIST OF CHOICE AND APPROVED VARIETIES,
ADAPTED TO THE CLIMATE OF THE UNITED STATES

TO WHICH ARE ADDED

FULL DIRECTIONS FOR THE TREATMENT OF

THE DAHLIA.

Illustrated by Engravings.

'———No flower that blows
s .ike the Rose, nor scatters such perfume."

NEW YORK:

C. M. SAXTON, BARKER & CO., 25 PARK ROW.

SAN FRANCISCO: H. H. BANCROFT & CO.

1860.

ADVERTISEMENT.

THE marked effect with which the "Cottage Bee Keeper" was received, as the first of the series of **"Saxton's Cottage and Farm Library,"** has led the Publisher to issue the present treatise on the Rose and the Dahlia.

No pains will be spared in bringing out the succeeding volumes agreeably to the plan of the original design; and if practicable, it is hoped that they will be still more deserving of success than those which have already appeared.

The publisher acknowledges his indebtedness to Messrs. Lea and Blanchard, of Philadelphia, for the permission granted for using Landreth's "Dictionary of Modern Gardening," in the compilation of this work. He also takes this occasion to thank Mr. William H. Starr for the use of several engravings herein employed.

C. M. SAXTON,

AGRICULTURAL BOOK PUBLISHER,

152 FULTON STREET.

INDEX.

THE ROSE.

THE DAHLIA.

THE ROSE.

INTRODUCTION.

Rose ! thou art the sweetest flower
That ever drank the amber shower ;
Rose ! thou art the fondest child
Of dimpled Spring, the wood nymph wild.
ANACREON.

THE Rose, the emblem of beauty and the pride of Flora, reigns Queen of the Flowers in every part of the globe ; and the bards of all nations and languages have sung its praises. Yet what poet has been able, or language sufficient, to do justice to a plant that has been denominated the Daughter of Heaven, the glory of spring, and the ornament of the earth ? As it is the most common of all that compose the garland of Flora, so it is the most delightful. Every country boasts of it, and every beholder admires it. Poets have celebrated its charms without exhausting its eulogium ; for its allurements increase upon a familiarity, and every fresh view presents new beauties, and gives additional delight. Hence it renovates the imagination of the bard, and the very name of the flower gives harmony to his numbers, as its odors give sweetness to the air.

To paint this universal emblem of delicate splendor in its own hues, the pencil should be dipped in the tints of Aurora, when arising amidst her aërial glory. Human art can neither color nor describe so fair a flower. Venus herself feels a rival in the Rose, whose beauty is composed of all that is exquisite and graceful. It has been made the symbol of sentiments as opposite as various. Piety seized it to decorate the temples, while Love expressed its tenderness by wreaths; and Jollity revelled adorned with crowns of roses. Grief strews it on the tomb, and Luxury spreads it on the couch. It is mingled with our tears, and spread in our gayest walks; in epitaphs, it expresses youthful modesty and chastity, while in the songs of the Bacchanalians their god is compared to this flower. The beauty of the morning is allegorically represented by it, and Aurora is depictured strewing roses before the chariot of Phœbus:

> "When morning paints the orient skies,
> Her fingers burn with roseate dyes."

The Rose is thought to have given name to the Holy Land where Solomon sang its praises, as Syria appears to be derived from *Suri*, a beautiful and delicate species of Rose, for which that country has always been famous; and hence called *Suristan*, or the "Land of Roses." The island of Rhodes owes its name to the prodigious quantity of roses which formerly grew upon its soil.

Of the birth of the Rose, it is related in fable, that Flora having found the corpse of a favorite Nymph, whose beauty of person was only surpassed by the purity of her heart and chastity of her mind, resolved to raise a plant from the precious remains of this daughter of the Dryads, for which purpose she begged the assistance of Venus and the Graces, as well as of all the deities that preside over gardens, to assist in the transformation of the Nymph into a flower, that was to be by them proclaimed Queen of all the vegetable beauties. The ceremony was attended by the Zephyrs, who cleared the atmosphere, in order that Apollo might bless the new-created progeny by his beams. Bacchus supplied rivers of nectar to nourish it, and Vertumnus poured his choicest perfumes over the plant. When the metamorphosis was complete, Pomona strewed her fruit over the young branches, which were then crowned by Flora with a diadem, that had been purposely prepared by the celestials to distinguish this Queen of flowers.

Anacreon's birth of the Rose stands thus translated by Moore:

" Oh ! whence could such a plant have sprung?
Attend—for thus the tale is sung;
When, humid from the silvery stream,
Venus appeared, in flushing hues,
Mellowed by ocean's briny dews —
When, in the starry courts above,
The pregnant brain of mighty Jove
Disclosed the nymph of azure glance—
The nymph who shakes the martial lance !
Then, then, in strange eventful hour,
The earth produced an infant flower,
Which sprung, with blushing tinctures drest,
And wantoned o'er its parent's breast.
The gods beheld this brilliant birth,
And hailed the rose—the boon of earth !
With nectar drops a ruby tide,
The sweetly-orient buds they dyed,
And bade them bloom, the flowers divine
Of him who sheds the teeming vine ;
And bade them on the spangled thorn
Expand their bosoms to the morn."

The first Rose ever seen was said to have been given by the god of love to Harpocrates, the god of silence, to engage him not to divulge the amours of his mother Venus; and from hence the ancients made it a symbol of silence, and it became a custom to place a Rose above their heads in their banqueting rooms, in order to banish restraint, as nothing there said would be repeated elsewhere; and from this practice originated the saying, *sub rosa,* (under the rose,) when anything was to be kept secret.

Oriana, when confined a prisoner in a lofty tower, threw a wet Rose to her lover to express her grief and love; and in the floral language of the East, the presenting a rose bud with thorns and leaves, is understood to express both fear and hope; and when returned, reversed, it signifies that one must neither entertain fear nor hope. If the thorns be taken off before it is returned, then it expresses that one has everything to hope; but if the leaves be stripped off, it gives the receiver to understand that he has everything to fear.

The Moss Rose is made the emblem of voluptuous love; and the creative imagination of a German poet thus pleasingly accounts for this Rose having clad itself in a mossy garment:

The angel of the flowers one day
Beneath a rose tree sleeping lay.

That spirit to whose charge is given
To bathe young buds in dews from heaven,
Awaking from his light repose,
The angel whispered to the rose,—
' Oh, fondest object of my care,
Still fairest found where all are fair,
For the sweet shade thou'st given to me,
Ask what thou wilt, 'tis granted thee.'
' Then,' said the rose, with deepened glow,
' On me another grace bestow.'
The spirit paused in silent thought—
What grace was there that flower had not?
'Twas but a moment—o'er the rose
A veil of moss the angel throws;
And, robed in nature's simplest weed,
Can there a flower that rose exceed?"

SPECIES AND VARIETIES

BOTANISTS enumerate at least eighty distinct species of the Rose, and it cists an almost innumerable number of varieties and sub-varieties, most of which are hardy, deciduous, or evergreen shrubs. To attempt a description or even to give a list of the names of all of these would be foreign to the design of this little treatise; and would be a needless waste of time, for the simple reason that many of them are unworthy of preservation, while in others, nothing short of the nicest and the most minute inspection can discover any difference.

The following are the names and characters of the more important and desirable members of this family, best adapted to this country and may be purchased at any of our principal florists:—

Bengal or Daily Roses.

Names.	Color and Character.
Animated,	Rosy blush.
Arsenie,	Light rose.
Augustine Hersent,	Superb rose.
Assuerus,	Crimson.
Admiral Duperre,	Dark rose.
Belle Isidore,	Crimson.

Names.	Color and Character.
Belle de Monza,	Dark rose.
Belle violet,	Violet purple.
Bisson,	Rosy blush.
Burette,	Dark red.
Cameleon,	Rose.
Cramoisi supérieur,	Crimson.
Cels,	Blush.
Comble de gloire,	Crimson.
Don Carlos,	Dark rose.
Duchess of Kent,	Pink.
Eugene Beauharnais,	Crimson.
Fabvier,	Scarlet.
Grandral,	Crimson.
Grandida,	Rose.
Hortensia,	Light rose.
Indica alba,	Pure white.
Jacksonia,	Bright red.
Louis-Philippe,	Crimson.
Lady Warrender,	White.
Laurencia,	Pink.
Marjolin,	Crimson.
Mrs. Bosanquet,	Large, blush.
Napoleon,	Rose, fine.
Reine de Lombardie,	Cherry red.
Samson,	Light rose.
Triomphant,	Crimson.
Vanilla,	Dark rose.

Tea-scented Roses.

Names.	Color and Character.
Archduchess Theresa,	White.
Aurora,	Blush.
Alba,	Pure white.
Arkinto,	Flesh color.
Adelaide,	Blush.
Antherose,	Blush white.
Adam,	Rosy blush.
Belle Marguerite,	Rosy purple.

1*

Names.	*Color and Character*
Bougère,	Light rose.
Boutrand,	Rosy blush.
Bon Silène,	Superb red.
Bourbon,	White.
Barbot,	Blush.
Camellia,	White.
Caroline,	Bright rose.
Countess Albemarle	Straw color
Duc d'Orléans,	Bright rose.
Devoniensis,	Creamy yellow.
Devaux,	Blush.
Delphine Gaudot,	White.
D'Arrance de Navarre,	Light pink.
Eliza Sauvage,	Pale sulphur.
Flon,	Buff.
Flavescens,	Yellow.
Golcondi,	Blush white.
Goubault,	Rosy blush.
Gigantesque de Lima,	Light yellow.
Gloria de Hardi,	Light rose.
Hyménée,	White.
Jaune Panaché,	Straw color.
La Sylphide,	Rosy buff.
Lilicina,	Lilac.
Lyonnais,	Rose.
La Pactole,	Yellow.
La Renomme,	White.
Madame Desprez,	White.
Mansais,	Rosy buff.
Niphetos,	White.
Odoratissima,	Rich blush.
Princesse Maria,	Blush.
Princesse d'Esterhazy,	Light rose.
Strombio,	White.
Triomphe de Luxembourg,	Rosy blush.
Victoria modeste,	Blush.
William Wallace,	Pale blush.

Bourbon Roses.

Names.	Color and Character.
Augustine Lelieur,	Bright rose.
Acidalie,	White, large, and fine.
Comte de Rambuteau,	Violet purple.
Ceres,	Dark rose.
Cytherea,	Rosy pink, very fragrant.
Comte d'Eu,	Bright carmine.
Doctor Rocques,	Purple crimson.
Dumont de Courset,	Deep purple.
Du Petit Thouars,	
Emilie Courtier,	Rosy red.
Gloire de Rosamene,	Brilliant crimson.
Gloire de Paris,	Bright red.
Grand Capitaine,	Brilliant scarlet.
Gloire de France,	Rose, very fragrant.
Hermosa,	Light pink.
Henri Plantier,	Pale rose.
Impératrice Josephine,	Creamy white.
Lady Canning,	Deep rose.
Madame Desprez,	Rosy lilac.
Madame Souchet,	Blush, fine.
Madame Lacharme,	Blush white.
Madame Nerard,	Light rose.
Maréchal de Villars,	Rosy purple, fine.
Ninon de l'Enclos,	Dark rose.
Paul Joseph,	Velvet crimson.
Princesse Clementine,	Deep rosy purple.
Phœnix,	Rose red.
Pierre de St. Cyr,	Light rose.
Queen,	Delicate blush.
Reine de Fontenay,	Brilliant rose.
Souchet,	Deep crimson.
Souvenir de la Malmaison,	Creamy white, fine.
Théresita,	Bright carmine.

Remontant or Hybrid Perpetual Roses.

In Europe, these Roses are highly esteemed; here, their reputation as "perpetuals," has been seriously injured, in consequence of their having been, in many instances, worked on stocks unsuited either to them, or to our climate.—*Landreth.*

Names.	Color and Character.
Antinous,	Dark crimson.
Aubernon,	Clear red, very fine.
Augustine Mouchelet,	Clear bright rose.
Baronne Provost,	Fine rose color.
Comte de Paris,	Dark crimson.
Claire du Chatelet,	Purple red.
Clementine syringe,	Pale rose.
Comtesse Duchatel,	
Crimson or rose du roi,	Light crimson.
D'Angers,	Delicate rose.
Doctor Marjolin,	
Duc d'Aumale,	
Duchesse de Nemours,	Pale rose.
Duchesse de Sutherland,	Bright rose.
Edouard Jesse,	Dark purple crimson.
Isaure,	Bright pink.
Israel,	Sable.
Insigne D'Estotells,	
Josephine Antoinette,	Rosy blush.
Louis Bonaparte,	
Lady Fordwich,	Deep Rose.
Lady Alice Peel,	Rosy carmine.
La Reine, or Queen,	Rose color, superb.
Madame Laffay,	Brilliant rose.
Marquise Bocella,	
Mrs. Elliott,	Rosy red.
Melanie cornu,	Deep crimson.
Newton,	
Palmyre,	Blush.
Princesse Héléne,	Large deep rose.
Prince Albert,	Very dark crimson, fine.
Prudence Rœser,	Rosy pink.

Names.	Color and Character.
Prince de Salm,	Dark crimson.
Prince of Wales,	Rose carmine.
Reine de la Guillotière,	Brilliant crimson.
Desquermus or Royal,	Large rose.
Stanwell,	Blush, very fine.
Sisley,	Large bright red.

Noisette or Cluster=Flowering Roses.

Names.	Color and Character.
*Alba,	Creamy white.
*Aimée Vibert,	Pure white.
Bengal Lee,	Blush, fragrant.
Cadot,	Blush lilac.
Charles Tenth,	Purple.
Conque de Venus,	White rose centre.
Cœur Jaune,	White yellow centre.
Champneyana,	Rosy white.
*Comtesse de Grillion,	Blush.
Chromotelle,	Large yellow, fine.
*Euphrosine,	Pale yellow.
Fellenberg,	Crimson, superb.
*Gabriel,	Blush, fine.
Jaune Desprez,	Rosy yellow.
*Julienne le Sourd,	Rose.
Julie de Loynes,	White.
Lamarque,	Creamy white, fine.
La Biche,	Flesh color.
Lady Byron,	Pink, fine.
Lutea, or Smithii,	Fine yellow.
Landreth's carmine,	Carmine.
*La Nymphe,	Pale rose.
Miss Simpson,	Blush.
Orloff,	Pink, fine.
*Ophire,	Yellow, fragrant.
Sir Walter Scott,	Deep rose.
Solfatare,	Superb dark yellow.
Vitellina,	White.

* Those marked * are dwarfs.

Climbing Roses.

These flower annually in immense clusters, grow rapidly, and are quite hardy.—*Landreth.*

Names.	*Color and Character.*
Banksia lutea,	Double yellow.
Banksia alba,	White.
Boursault,	Rose color.
Boursault purpurea,	Purple.
Boursault blush,	Large blush.
Boursault gracilis,	Bright rose.
Bengalensis scandens,	Large rosy white.
Félicité perpetuelle,	Blush white.
Grevillia,	Greville produces immense clusters, of various colors and shades, from white to crimson.
Multiflora,	Pink.
Multiflora alba,	Blush white.
Rubifolia, Single Michigan or Prairie,	
Rubifolia elegans,	Double pink.
Rubifolia purpurea,	Double purple.
Rubifolia, Queen of the Prairies,	Double pink.
Rubifolia alba,	Double blush white.
Russelliana,	Crimson cottage rosa.
Sempervirens plena,	Superb white.
Triomphe de Bollwyler,	Blush white.
Laura Davoust,	White.

Microphylla Roses.

Names.	*Color and Character*
Maria Leonida,	White, extra fine.
Microphylla rosea,	Rose color.
Microphylla odorata alba,	Creamy white.

Musk=scented Roses.

Names.	Color and Character.
Moschata,	White semi-double.
Moschata superba,	Pure white, very double.
Princesse de Nassau,	White double.

Hardy Garden Roses.

Names.	Color and Character.
Miaulis,	Rosy purple.
Coronation,	Purple crimson.
Reine des roses,	Bright crimson.
Duc d'Orléans,	Dark rose.
Painted Damask,	White.
Brennes,	Dark pink.
Rivers' George IV.,	Superb crimson.
Hybride blanche,	White.
Heureuse surprise,	Carmine.
Ranunculus,	Purple, compact.
La capricieuse,	Purple crimson.
Royal Provins,	Superb pink.
Du Roi,	Perpetual, bright red.
Harrisonii,	Yellow Austrian brier.
Moss, Single,	Crimson, very mossy.
Moss, Common,	Rose.
Moss, Luxembourg,	Crimson.
Moss, White,	Perpetual.
Moss, Crested,	
Moss, Adelaide,	
York and Lancaster,	Red and white.
Provins Belgic,	Large pink.
Four Seasons,	Pink.
Moretti,	Light rose.
Burgundy,	Rose, compact.
Persian,	Double yellow.
Village Maid, or La Belle Villageoise,	Rose, striped with lilac.
Austrian Brier	Deep yellow

In contemplating some cf the best Roses from the various families, we cannot help admitting, that, compared with the old and still valued varieties, more than two-thirds even of our selections are not so good in character. The love of novelty is all-powerful; a shade of color, the slightest difference in habit, a different season of bloom, an alteration in the size or color of the foliage, the distinction between a slow and a fast growth, have always been considered sufficient by sellers to warrant a new name and a place in the catalogues; and the Rose, unlike all other flowers, began with better varieties than hundreds of their successors, or rather their younger rivals, proved to be.

Notwithstanding many of the early Roses were really beautiful, and hardly admitted of much improvement, we had, at a very early period of the fancy, such Roses as the Tuscan, the Cabbage, the Cabbage Moss, the Maiden's Blush, White Provence, and Double Yellow. These have, it is true, been succeeded by a few worthy of ranking with them, but they have to be selected from thousands infinitely worse, and hundreds which ought not, for the raiser's honesty, or the buyer's good sense, to have even passed the seed bed. If, therefore, we were to select, to lessen our readers' difficulty in choosing, we could not recommend them as Roses equal to old favorites; for not one in fifty would beat the few we have mentioned, and which ought to be the first they furnish.

The Provence Rose.

The Provence Rose, or, as it has been called, the Hundred-leaved Rose, is a distinguishing title for every Rose that has a remarkably double flower, unless there is something in the habit or character that claims for it another title. If this were understood, we should know what we are about. The Moss Rose would clearly come under this, were it not for the moss; for the old Cabbage Rose, and the Moss Rose strongly grown, would not be known from each other, except for the Moss; and the Moss Rose would be a Moss Rose, if ever so single, though its original were double and fine. Now, the Provences, of which the old Cabbage Rose is a sort of type, and generally called the Hundred-leaved Rose, ceases to deserve this name, if semi-double. So that although the origin of the family is rightly named, many pushed into the same list do not deserve the name.

Moss Rose.

This family is distinguished by the mossy appearance of their stems and the calyx, and therefore there is no difficulty in recognising any member of the family.

The French Rose.

This, to some of our readers, would appear to mean roses raised in France. It happens, however, that the original was, as many of the leading ones were, raised by Van Eden, in Holland, and it was years before the French raised a single seedling from them; nevertheless some of the so-called varieties were raised in France, but as there are hundreds raised in that country which are not belonging to this family, the distinguishing name fails; and were it not so, they are so unlike each other that one could not recognise, in any particular feature, enough to decide, nor do the rose growers themselves appear more certain.

Hybrid Provence Roses.

These are said to be intermediate between French and Provence roses, because they have the long shoots of one and the dense foliage of the other; the said long shoots and dense foliage being the characteristics of roses of other families in quite as large a degree, and even in this very family, we have varieties which seem to be between the Boursault and Provence. So that all is indecision, change, uncertainty, and frivolity. In this family, the distinguishing character is that they "are robust and hardy;" so are hundreds that do not belong to it.

Hybrid China Roses.

We are told of this family, that the numerous varieties give a combination of all that is beautiful in a Rose. They are said to owe their origin to all sorts of crosses; but there is a distinguishing feature in these, if it be adhered to: "leaves smooth, glossy, and sub-evergreen; branches long, luxuriant, and flexible." Then, again, we are informed "that hybrids produced from the Rose, impregnated with the China

Rose, are not of such robust and vigorous habits as when the China Rose is the female parent." This looks like plain, straightforward information; but it is followed by the same incertitude as some of the other distinguishing features of families. Mr. Rivers adds: "But, per haps, this is an opinion not borne out by facts; for the exceptions are numerous, and like many other variations in roses, and plants in general, *seems to bid defiance to systematic rules.*" Of course, they do; and, with the exception of those names which bespeak a distinct character, the splitting of this beautiful flower into so many different families at all, was a very injudicious measure. Athelin, a Rose classed in this group, is called also a Hybrid Bourbon, and as it blooms in clusters, would have been much better understood if called a Noisette. It comprises other roses as unlike each other as can be well imagined, and many of them will shoot ten feet in a season, and would be much more at home if classed as Climbing Roses. Belle de Rosny, among this family, is nevertheless called also a Hybrid Bourbon, and many others of this family are destined to be removed, if the senseless dis tinctions by name are to be kept up.

White Roses.

Here we have an illustration of the extreme folly of the present distinctions. We are told the roses of this division may be easily distinguished by their green shoots, and leaves of a glaucous green, looking as if they were covered with a grayish impalpable powder; and flowers generally of the most delicate colors, graduating from a pure white to a bright but delicate pink.

The Damask Rose.

. .us is as incongruous a group as any. Blanche borde de rouge has flowers sometimes a pure white, at others margined with red. Claudine has flowers of a pale rose color. York and Lancaster, also classed among them, has flowers striped with red and white. Coralie is flesh color. Then we have Madame Hardy, which, we are fairly told, "is not a pure Damask Rose;" perhaps not, as it is white, and unlike all the rest. Then, there is the Duke of Cambridge, which Mr.

Rivers "at first thought a Hybrid China," and says, "will, perhaps, be better grouped with the Damask Roses.'

Scotch Roses.

So long as this family was allowed to be kept select, these roses were very distinct; they make long briery shoots, and flower with small blooms almost like briers, the whole length of stems. They are exceedingly pretty, formed as a bank, or in clumps. They are not adapted for standards. They bloom early, and the Scotch nurserymen now boast of two or three hundred varieties; but like all the other families, there are many among them that have been raised from seed, and others imported, which are neither by name nor nature Scotch. Amiable étrangère is a French hybrid. Adelaide is a large Red Double Rose. La Cenomane is a French hybrid with large flowers, "not so robust as the pure Scotch varieties."

The Sweet Brier.

This lovely ornament, or rather tenant of the garden, is universally admired for the delicious fragrance of its foliage, and for nothing else. It is only necessary to say here, that others whose leaves are not fragrant have been placed with it to make a family; some of the new members having but little fragrance, and one, the Scarlet Sweet Brier, none at all.

The Austrian Brier.

Here we have the same evidence of indecision as to where things ought to be placed. In this scentless family we have Williams' Double Yellow Sweet Brier. In fact, the Sweet Brier and the Austrian Brier are muddled together so completely that catalogues do not agree, and the further we go, the more confusion we get into, and more instances occur of removal from one division to another.

The Double Yellow Rose

Here we have only two individuals, the old Double Golden Yellow, sc beautiful and double as to be universally admired, and the Jaune, a dwarf kind, both shy bloomers under ordinary management, or when we come to the right of it, never blooming well till they are matured, which takes some years. Of course, there are many Double Yellow Roses, but only two are admitted into this select family.

Climbing Roses.

Here we might expect to find all those roses which, from their habits, were adapted to the fronts of houses, pillars, trellises, and other lofty stations. One would, at least, expect that, if Climbing Roses mean anything, it means all roses that will climb. No such thing. Having pushed, we know not how many roses that climb into other families, of course they cannot be here. We have various divisions in this family notwithstanding: First, we have the Ayrshire Rose, which is said to be a hybrid, accompanied by several others called Ayrshire Roses also; next, we have the second division, called Rosa multiflora, said to be a native of Japan, and a number of companions as unlike it as may be; not that there are any among this family that do not climb, but there are very many as good Climbing Roses shut out from it.

The *Queen of the Prairies*, or Michigan Rose, is remarkable for its perfectly hardy growth, flourishing equally well in Canada at the north, and in Texas at the south. It grows with unparalleled rapidity, exceeding all other roses of this family, covering an entire arbor or an old building in a short space of time. It blooms, also, after other summer roses are mostly gone, its flowers occurring in large clusters of different shades.

Evergreen Roses.

Here there can be no mistake: an Evergreen Rose must be an Evergreen Rose; but, although we have some enumerated, there are plenty of Evergreen Roses not admitted into this family, but pushed about in all directions, some crammed into the China, and some into the Hybrid China.

Boursault Roses.

This is said to be "a most distinct group of roses, with long reddish flexible shoots;" yet Gracilis is affirmed to be "unlike the other varieties of this division." They are said to be good Climbing Roses, making ten feet of growth in the season.

Banksian Roses.

The White and Yellow Banksian Roses are very beautiful plants, with small foliage and flowers, very graceful, and distinct as any in cultivation; yet we have a rose-colored hybrid introduced with them; a plant acknowledged to partake "as much of the character of the Boursault Rose, as of the Banksian."

Hybrid Climbing Roses.

These, one would think, are neither Climbing nor Dwarf, but between both. Not so, however; because Rosa craculum makes shoots from ten to fifteen feet in a season. Madame d'Arblay, or Well's White, has been formerly placed among the Evergreen Roses; but whether she misbehaved herself there, or was a great favorite here, is of no consequence. She was removed from that family to this. We are, however, informed, with regard to her sojourn among the family of Evergreens, and subsequent removal, that her "habit is so different and her origin so well ascertained, that Mr. Rivers removed her to the present family."

Perpetual Roses.

These, if the rose gentlemen would stick to the character, would be very easily defined—roses which have a complete season of bloom; which go off but a short time; make a fresh season of bloom, and so on. Not like the China Roses, always "growing and blooming," but fairly making different seasons of bloom, as complete as if a winter intervened.

The Bourbon Rose.

The original Bourbon Rose was a hybrid between the Common China and the Red Four Seasons. Of course, this was quite enough reason for rose growers to add to the family all that were something like it, and others that were nothing like it. Here let Mr. Rivers speak: "Diaphane is a small high-colored Rose, almost scarlet. This is not a true Bourbon." The fact is, there is nothing like the Bourbon Rose about it. Here we have also Gloire de Rosamène, unlike the Bourbon Rose in everything. It is a robust Climbing Rose, of which even Mr. Rivers himself says, " As a Pillar Rose, it will form a splendid object." The White Bourbon, which the French cultivators are at war about, "some swearing," as Mr. Rivers tells us, "by all their saints that it is a veritable Bourbon, while others as strongly maintain that it is a Noisette;" and from its clustered flowers the latter are nearest right. But all this arises from the multiplication of families.

China Roses.

Everybody knows the Pale China and the Dark China Roses, which may be seen decorating the cottages of our industrious classes as well as the gardens of the rich. They were, however, Bengal Roses, and not natives of China. Now the distinguishing characteristic of the Bengal, or, as now called, China Rose, is smooth bark, with the thorns distant from each other; shining leaves, and constant growing and blooming. These features could be well understood by everybody; but everything that can be at all traced to have any one of these features, and cannot be easily placed in other families, must come to this; and so we have plenty, and a most beautiful family it is.

Tea-scented China Roses.

This is an acknowledged variation of the Bengal, or, as the rose dealers will have it, China Rose; but it is a true China, imported into England from that empire in 1810. It is said to have been the parent of this large family; but here we have the same difficulty that presents itself in other families—there is no place to draw the line; they are China Roses, and only China Roses, but they are stronger scented than the Bengal, called Common China, and it is difficult to detect the difference between the highest perfumed of the former class and the lowest perfumed of the China Tea Roses, as now classed.

Miniature Roses.

This family is also said to be China, possessing all the marked features; but it is smaller than the others, and is acknowledged by Mr. Rivers to be only a dwarf variety of the Common China, or, as we insist, Bengal. It is worthy of remark, that all those so-called China Roses have the characteristics we have mentioned, the constant grow-

!ng and blooming, if kept in order under proper protection; and are not deciduous.

The Noisette Rose.

The distinguishing character of this Rose is that it flowers in bunches, and this ought to be the character of every one added to the family. But here we have Lamarque, which is anything but a Noisette; it does not flower in bunches, unless every Rose which has two or three flowers on a stem is to be called Noisette; and Smith's Yellow Noisette is about as much entitled to the name of Lamarque. But they are not alone; too many which have no claim on the family have nevertheless been forced on them.

The Musk Rose

This is an old favorite, and many which nave been supposed to come from its seed are fastened on it as a family, and many not very like the parent. The family, like some of the others, is greatly confused, and there is nothing so distinct as to connect it as a separate class.

The Macartney Rose.

The characteristic of this Rose is its very bright thick evergreen foliage, and therefore any other Hybrid Roses which have that characteristic might, according to other classifications, be put among these. Maria Leonida is perhaps the best of them; Rosa berberifolia hardii, of whose origin Mr. Rivers makes a sad muddle, is classed with this family. Mr. Rivers' story is, that " Rosa hardii was raised from seed by Mons. Hardy, of the Luxembourg Gardens, from Rosa involucre, a variety of Rosa bractcata, fertilised with that unique rose, Rosa berberifolia which was very frequently exported from Persia, and comes always true to the parent; some of the Persian seed was sent to Mons. Hardy, and from that he, like others, raised the true Rosa berberifolia, which Mr. Lee, of the Hammersmith Nursery, raised from Persian seed likewise, more than twenty years before Mr. Hardy was a rose raiser at all." Well may Mr. Rivers say, in continuation, "This curious hybrid, like its Persian parent, has single yellow flowers, with a dark eye, and evergreen foliage." The fertilising part of the busi

ness is the mere work of a fertile imagination. When any one has
got Rosa berberifolia, he need not trouble himself about whether he
has it from the seed raised by Mons. Hardy, or the seed raised by his
predecessors. There is no more variation, and no more hybrid about
either, than there is in two plants of small salad.

Rosa Microphylla.

  This, we are told, is nearly allied to the Macartney Rose; so are
the varieties of it, and ought not to have been separated.

QUEEN (ROSA bourboniana).

JAUNE DESPREZ (Yellow Rose)

YELLOW BANKSIAN ROSE (Rosa banksia lutea

SMALL LEAFLETTED ROSE (Rosa microphylla).

BURGUNDY ROSE (Rosa gallica).

THE VILLAGE MAID (La Belle Villageoise).

WILLIAMS DOUBLE-YELLOW SWEET BRIER.

CHARACTERISTICS OF A FINE ROSE.

THERE is no flower more difficult to define than the Rose, and the difficulty arises out of several curious facts. First, it is the only flower that is beautiful in all its stages—from the instant the calyx bursts and shows a streak of the corolla, till it is in full bloom. Secondly, it is the only one that is really rich in its confusion, or that is not the less elegant for the total absence of all uniformity and order. The very fact of its being beautiful from the moment the calyx bursts, makes the single and semi-double roses, up to a certain stage, as good as the perfectly double ones; and there is yet another point in the formation of some varieties, which makes them lose their beauty when they are full blown. For instance, the Moss Rose is a magnificent object so long as the calyx is all seen, but so soon as the flower fully expands, all the distinction between a Moss Rose and a common one has departed, or is concealed. This brings us at once to an acknowledgment that the grand characteristic of a Moss Rose is its calyx. These properties must never be estimated by full-blown flowers, and therefore, all varieties of Moss Roses must be exhibited before they expand enough to hide the calyx.

There are some properties, however, which apply to all roses, whatever be their characteristics in other respects, and, therefore, must be taken as an estimable point in the construction of a flower.

1. The petals should be thick, broad, and smooth at the edges.

Whether this be for a Moss, which is never to be shown fully opened, or the florist's favorite, which is to be shown as a dahlia, this property is equally valuable, be use the thicker the petal, the longer it is opening, and the longer does it co tinue in perfection, when it is opened. There is another essential point gained in thick-petalled flowers: The thicker the petal, the more dense and decided the shade or color, or the more pure a white, while the most brilliant scarlet would look tame and watery if the petal were thin, transparent, and flimsy. Hence, many semi-double varieties, with these petals, look bright enough while the petals are crowded in the bud, but are watery and tame when opened, and dependent on their single thickness.

2. The flower should be highly perfumed, or, as the dealers call it, fragrant.

Whether this is to climb the front of a house, bloom on the ground, or mount poles or other devices, fragrance is one of the great charms. which place the Rose on the throne of the garden as the Queen of Flowers.

3. The flower should be double to the centre, high on the crown, round in the outline, and regular in the disposition of the petals.

This would seem to be a little contradictory, after saying, that in a Moss Rose, the full-blown flower cannot be allowed, because it conceals the grand characteristic of the plant. But it is not contradictory, because we defend it on grounds which render doubleness equally valuable to the Moss family, which should not be shown in full bloom, as to those which are so exhibited. The more double the flower, even when amounting to confusion, the more full and beautiful the bud in all its stages. Those who have noticed the single and semi-double Moss Roses will remember that the buds are thin and pointed, and starved-looking affairs, while the old common Moss Rose, which is large and double as the Cabbage Rose, is bold, full, rich, and effective, from the instant the calyx bursts. At this point, we shall have to branch off and take families; perhaps the Moss Rose family is the best to commence with. Those who now follow through the different species or varieties, will find the first three rules are essential to all, and are therefore repeated with each division.

Properties of Moss Roses.

1. The petals should be thick, broad, and smooth at the edges.

2. The flower should be highly perfumed, or, as the dealers call it, fragrant.

3. The flower should be double to the centre, high on the crown, round in the outline, and regular in the disposition of the petals.

4. The quantity of moss, the length of the spines, or prickles, which form it, and its thickness, or closeness, on the stems, leaves, and calyx, cannot be too great.

This being the distinguishing characteristic of Moss Roses, the more strongly it is developed the better.

5. The length of the divisions of the calyx, and the ramifications at the end, cannot be too great. As the entire beauty is in the unde-

veloped bud. the more the calyx projects beyond the opening flower, or rather the more space it covers, the better.

6. The plant should be bushy, the foliage strong, the flowers abundant and not crowded, and the bloom well out of the foliage.

DIAGRAM OF A FINE DOUBLE ROSE.

7. The color should be bright or dense, as the case may be, and if the color or shade be new, it will be more valuable; and the color must be the same at the back as the front of the petals.

These seven properties would constitute a Moss Rose a valuable acquisition, and probably, at present, the greatest acquisition would be a yellow one.

8. The stem should be strong and elastic, the footstalks stiff, so as to hold the flower well up to view.

2*

Properties of Roses for Stands, showing the Single Bloom like Dahlias.

1. The petals should be thick, broad, and smooth at the edges.
2. The flower should be highly perfumed, or, as the dealers call it, fragrant.
3. The flower should be double to the centre, high on the crown, round in the outline, and regular in the disposition of the petals.
4. The petals should be imbricated, and in distinct rows, whether they be reflexed, like some of the velvety Tuscan kind, or cupped like a ranunculus; and the petals to the centre should continue the same form, and only be reduced in size.
5. The color should be distinct and new, and stand fast against the sun and air, till the bloom fail.
6. The stem should be strong, the footstalk stiff and elastic; the blooms well out beyond the foliage, and not in each other's way.

The very worst habit a Rose can have, is that of throwing up several blooms close together, on short stiff footstalks, some of which must be cut away before the others can be fully developed; as show flowers, they are bad, and as plants, they are very untidy. The side buds prevent the centre flowers from opening circularly, and when the first beauty is off, they exhibit dead roses held fast between two living ones.

Properties of Noisette Roses.

However singularly some catalogues class these varieties, we intend, by this name, to distinguish those roses which bloom in clusters.

1. The petals should be thick, broad, and smooth at the edges.
2. The flower should be highly perfumed, or, as the dealers call it, fragrant.
3. The flower should be double to the centre, high on the crown, round in the outline, and regular in the disposition of the petals.
4. The cluster should be sufficiently open to enable all the flowers to bloom freely, and the stems and footstalks should be firm and elastic, to hold the flower face upward, or face outward, and not hang down, and show the outside, instead of the inside of the blooms.
5. The bloom should be abundant at the end of every shoot.

6. The blooming shoots should not 'exceed twelve inches before they flower.

˙. The bloom should stand o .t beyond the foliage, and the plant should be compact and bushy.

We now proceed to a family which we shall designate Climbing Roses, and which comprise blooms of the Noisette kind, that is, in bunches; blooms which come singly, large and small; flowers early and late; and, in fact, which comprise all sorts of roses that grow tall enough for training.

Properties of Climbing Roses.

1. The petals should be thick, broad, and smooth at the edges, with the outer ones curving slightly inwards.

2. The flower should be highly perfumed, or, as the dealers call it, fragrant.

3. The flower should be double to the centre, high on the crown, round in the outline, and regular in the disposition of the petals.

4. The joints should be short from leaf to leaf. The blooms should come on very short branches, and all up the main shoots. The plant should be always growing and developing its flowers, from spring to autumn, and the foliage should completely hide all the stems, whether the plan be on front of a house or on any given device.

Concluding Remarks.

Having now travelled through the chief of the families, which require separate notices of their properties, the first three properties numbered being required in all of them, we add, by way of a finish for all, except Moss Roses, that

The foliage should be bright green and shining, and, though not likely to be found in many varieties, it should be permanent, and constitute an evergreen.

By this, we mainly establish a point in favor of an evergreen. We mention nothing about size, because size forms the distinction between many roses which have no other difference, and has little or nothing to do with the properties of the Rose, except uniformity in the same variety.

PROPER SOIL FOR THE ROSE.

The proper soil for the Rose is strong rich loam, and well decomposed vegetable mould, cow dung, or horse dung; but as we are
too often already provided with the kind of soil we are obliged to use
and the gardens and situations for our roses are generally ready
made, all we can do is to modify and supply the deficiency, if any,
as well as we can. If the soil be light, holes must be dug, and loam
and dung forked in at the bottom of the hole, as well as the hole be
filled up with the same mixture; for troublesome as this may be, it is
the only way to secure a good growth and bloom, and it is next to
useless to plant roses in poor light soil without this precaution.
Kitchen gardens well kept up, will always grow the Rose well, and
unless the soil be very poor and very light, a good spadeful of rotten
dung, mixed with the soil where the Rose is planted, will answer all
the purpose. Among the evils of poor soil for the Rose, it is not the
least, that it frequently makes the flower that would otherwise be
double come single or semi-double, so as to destroy all identity of the
variety by its bloom; and although many thousands of roses of no
value have been sent out, many others which did not deserve it have
been condemned, because the party who was growing them knew
nothing about their cultivation, and starved them into a false character. As it is difficult, however, to give the Rose too rich a soil, it
may be as well, even if you think it good enough, to work in a spadeful of dung with it; for it will do no harm, even if the state of the
ground be ever so good. We have no doubt that the Rose would
flourish in rotten turf, and when they are to be grown in pots, it is
practicable to give them this invaluable stuff to grow in; but unless
it be a recently turned-up pasture, there is nothing approximating to
it out of doors, and even this is far less supplied with the rotted grass,
than when turfs are cut thin to rot for use. As a general principle,
then, may be laid down that the Rose requires rich soil; and that if
you have it not, you must change the nature of what you have, by
means of dung, or loam, or both

MANURES FOR THE ROSE.

ONE of the best manures for the rose is a mixture of one part of Peruvian guano, three parts charred turf and earth, and six parts of cow dung. A thin dressing of this should be pointed in with a trowel every spring.

Roses may also be watered at any period of their growth with a mixture of one fourth of a pound of Peruvian guano and eight gallons of water, to be applied with a watering pot in the evening or on a cloudy day.

PLANTING OF THE ROSE.

To plant the rose properly, the root must first be examined, and every particle of it that has been bruised should be cut off with a sharp knife just above the bruise; all the torn and ragged ends should be made smooth, and cut away as far as they are split or damaged. If any root has been growing downward, it should be shortened up; for it is better to discourage any from growing downright. This preparation being made, and the holes dug large enough to take the root in without cramping it, fork or dig up the bottom of the hole to loosen it, and, if necessary to make any addition to the present soil, to mix it properly with the soil taken out, and work it some way into the soil at the bottom. Let one hold the tree or plant, if it be too large to manage properly alone, and the other throw in the soil between the roots. By moving the stem backward and forward, and pulling upward a little, it is easy to work the soil well between the roots, and on this much depends. When it is adjusted, the top of the root must be pretty close to the top of the ground; there must be none of the stump or stem buried; and when trodden down, the root must be fixed steady and solid. If you have to manage the planting by yourself, you must, as soon as the hole is prepared, lay hold of the stem just above the root, and return the soil with your other hand, continuing to move the head first one way and then the other, until

the soil has worked well between the roots, when it may be trodden
in as mentioned before.   ·

Dwarf plants there is no difficulty in planting, but you must be
careful to keep the crown of the root near the surface of the ground,
the treading in of all fair and solid being a necessary operation with all
the kinds of plants. With the standard sorts you should drive stakes
into the ground pretty firmly, and fasten the stems of the roses to
them, to prevent the wind from removing them; as when your roots
have been once firmly trodden in, you cannot move a tree one way
nor the other without breaking the fine fibres, and thus lessening the
capacity of the root to carry strength to the head. If you are planting a group of standard roses, you should place the highest in the centre, and the lower ones nearer the outside; in fact, a handsome clump
of roses might have six-foot standards in the middle, four feet six inches
in the next row, three-foot ones nearer the front, and eighteen-inch
ones outside; these, if at proper distances, and with picked sorts, of
something near the same habit of growth, will form a superb mountain
of roses in the proper season.

Rows of Standard Roses may be planted with advantage on each
side of a coach road, in a park, or on both sides of a path on a lawn,
but at proper distances, so that each shall form a specific object in
itself, as well as a portion of a row of rose trees. Roses also form very
beautiful objects planted in isolated situations on lawns, and especially
when the sort of rose is distinct from others, or blooms at different
periods; for whatever forms a portion should be of a similar habit to
the rest of the whole. Thus, if a particular walk in a garden or shrub ·
bery were bounded by two rows of roses, they should all flower at
once. If a clump of roses is planted, they should flower at one season.
A mixture of spring, summer, and autumn roses would be very bad ;
the place never looks right; therefore some pains must be taken to
keep all those which flower at the same period of the year together.
One portion of the garden may then be always garnished with roses,
and it is far better than having them straggling about, with here and
there a flowerless one among those in bloom, or a blooming one among
those not in flower.

Planting of roses which are on their own bottoms, or worked low
down for dwarfs, or for climbers where flowering wood is always
vanted from the ground, differs in nowise from any other planting

except as to the situation, which should be chosen not too much exposed to the wind, as in the most sheltered spot they always have enough to encounter. They must be planted firmly, and in good soil; and whatever they have to climb up should be firmly placed by rights before they are planted, but certainly before .they shall have grown much, as the roots spread a good deal, and if damaged by violence after they have begun to grow vigorously, they will receive a check which they may not get over the same season.

POTTING OF ROSES.

To the cultivators of the Rose, any improvement in pots is of importance. Those designed to grace a hall or a window of a dwelling, may be made in fine stone and earthenware of various patterns, and should be so constructed as to possess advantages over the common old red porous ones made of clay. One reason why plants potted the usual way do not flourish well in the house during the winter season is, the proper want of leakage, or drainage, and a due circulation of air about their roots, in consequence of the close connection between the bottom of the pot and the shelf or bench on which it rests.

Mr. M'Intosh, gardener of the Duke of Buccleuch, has obviated the above-named objection by making pots with feet, as denoted in the adjoining cut. By this means, the plants get rid of their moisture, and freely receive air about their roots through the hole in the bottom of the pot.

Potting Deciduous Roses for Forcing.

The nearer you can imitate planting in the open ground the better. The soil should be the same or richer, with dung chiefly, because you cannot water soil without washing away, in some measure, whatever it is impregnated with, that is soluble. By a parity of reasoning, you cannot moisten with water impregnated with anything, without imparting the virtue or mischief of the solution to the soil

It is the best way to use half of rotted turf and half of rotted da $\sim\hspace{-2pt}g$;
if it be not too light to let water pass freely, add a little turfy peat,
broken through a sieve that would pass a hazel nut. Trim the roots,
to get rid of all bruises; and, in the first instance, choose plants, the
roots of which are within a moderate compass, for pot culture, and
are well taken up. Select pots that will receive the roots without
much cramping; carefully put the soil between and among the fibres
and larger roots; strike the pots on the potting table, and poke the
soil down so as to be firm.

If the roses be dwarf, follow the directions about pruning at once,
and let them be placed in a cold frame, watered, to settle the earth
about them, and covered up. This should be done in the Southern
and Middle States from November to February, when those for forcing should be put into the greenhouse, gently increased in temperature, well watered, and kept growing hard; any buds that show
should be removed, and they should be allowed to complete their
growth, and then be plunged in the open ground, and there the wood
be permitted to ripen. When the leaves have fallen, and the wood is
fairly ripe, they may be pruned, by removing all the weak shoots, and
shortening the strong ones; the balls turned out to examine, and if
matted with roots, pots a size larger be given. They may then be
placed in a cold frame, plunged to their rims, until the period you
want to force them. They will flower better the second year than
they could have flowered the first, and if the blooms are all picked
off again as fast as they show, instead of being allowed to perfect
themselves, the growth will be more free; and by growing hard to
complete it early, and leaving them out again to ripen, they will allow
of being pruned into a handsome form, being carried into the house
sooner, and will flower most abundantly, instead of having one or two
sickly shoots with their miserable half-starved blooms. At the end,
they will have as many as you please to leave eyes for, pruning them
the same as you would standards or bushes out of doors, and the
blooms will come as rich, as handsome, and as well colored as any in
the open air. Roses may then be forced at almost any season, only
they ought to undergo the same forcing a season or two without
being allowed to flower, that they undergo the season they are to be
forced into bloom. And this will answer season after season when
they are once well established, for they require only the usual shifts

of plants, which have their balls matted with root; but of the forcing, more hereafter.

Potting for Show.

As it is at length the fashion to show roses in pots, the only proper plan of showing any but single blooms, face upward, the plan of potting cannot differ from those potted for forcing. Presuming that if they are late roses and require forcing, they will be treated after the plan above mentioned, so far as the potting is concerned, the difference between what the perfectly hardy and summer or autumn blooming roses will require after potting, as we have directed, is to be put out in an open situation; and if standards, they should be fastened to a railing, or trellis, as well as being plunged in their pots, that the wind may not disturb them. Here they may be protected various ways: a mat thrown over the head of a rose protected it, though not a very hardy one, against the last winter's frost. A wisp of straw tied at one end, and opened cap-like over each and among the branches of roses, protected them a good deal, and probably, had they not been autumn pruned, might have protected them entirely from mischief, but as it was, some of the pruned branches died back, though the unpruned ones did not.

Potting the Small, the Smooth Wooden, and Chinese Varieties.

Here, from the first, the soil should be one third rotted dung, one third peat, and one third the loam of rotten turf. In this stuff, the most delicate will succeed. From the period of their having struck root, they can hardly do wrong if potted in this soil, in a proper-sized pot, with ordinary drainage. Small plants should be placed in pots no larger than the roots require to hold them, with a moderate share of earth to live in. This kind of rose should be kept growing in a cool frame or greenhouse, or pit, with not much moisture; plenty of air in dry mild days, and a refreshing shower when it is warm. It is safer to plunge them in ashes, if you can, up to the rims of their pots: it keeps them moist longer than if the pot is exposed, it mostly does, in bad weather; and though it perhaps does not kill them, it makes them weakly for some time. In this way, they may grow from time to time, and be shifted from one sized pot to another, requiring only

that the buds should be plucked off directly they show, sc .ong as the
plant is wanted to grow fast.

FORCING OF EARLY ROSES.

. THIS art consists in bringing the Rose, by. degrees, out of its season,
as we have half explained under the head of "Potting for Forcing."
We know that a Rose can be potted in January, and made to produce
flowers in May; but those who wish to force should know the best
way.

A Rose, then, for early forcing, requires three seasons to be perfect. The first season, it should be put into a greenhouse, and
from thence into the stove, as early as November. It is sure to grow,
no matter what sort it is; and let it grow its best, but pluck off the
buds if it have any, yet it should not be drawn; this can be managed
two or three ways, but it requires, to prevent drawing, light and air.
These will have grown pretty well as large as they can grow, by the
time they may be turned out and plunged in the open air. The wood
will ripen well in the summer time; and in October, re-pot them into
a size larger pots; prune them by taking off all the weak shoots, and
all the least valuable of those in each other's way; shorten the best
wood to two or three eyes, thinning the inner branches all that may be
necessary to give air, light and freedom to the new wood. Take them
into the greenhouse, thence, soon, into the stove. Let the bloom
buds, as they appear, be plucked off, and the growth to be perfected
again, which will be earlier than the previous season, as they were set
growing earlier. Be early in your attendance on them, when they
commence growing, so as to remove useless buds, instead of allowing
them to form useless branches. When the growth is completed, remove them into a cold frame, to be kept from the spring frosts, but
where they can have all the fine weather. In this state, they may
remain till they can safely be put out in the open air, plunged into the
ground, and properly fastened to protect them from the wind. In
September, you may examine the balls of earth, to see if the roots have
room; if c atted at all, give them another change. Prune the plants

well, as before removing altogether such of the present year's shoots
as are at all weakly, and shortening all the best to two or three eyes.
Let them now be taken to the greenhouse, or conservatory, or a
grapery, or all in turn; but gradually increase the temperature, till, by
the end of October, they may go into the forcing house, beginning at
the temperature the house was that they came from, say fifty to fifty-
five, and continuing it till they are fairly growing; then increasing it
to sixty, and eventually to sixty-five; rubbing off, as before, all useless
shoots, and giving plenty of air, when it can be done without lowering
the temperature. At the least appearance of the green fly, syringe with
plain water; fumigate at night, for too strong a smoke would all but
destroy the plants and incipient blooms. In this way, you will be clear
of the pest without damage, and your reward will be a fine show of
blooms on every rose tree; strong growth, healthy foliage, handsome
plants, and all that can be desired.

Forcing Later Roses.

The principle on which the early forcing is conducted must be carried
out in full, not only in potting the plants then pruning, but also in the
period of removing them. If you wish those a month later to succeed
the first, put them into the house a month later, each of the years.
If you want others to succeed these second, put them into the house
a month later still each year. For nothing has been shown yet in the
way of pot roses, better than were shown several years ago, and all of
them have had a weakly drawn appearance, and have been anything
but creditable to the taste of the gardeners; for they have been staked
all over, and thin, flimsy roses on limp-lankey stems, bound up to a
thicket of unnatural wood. Now, by the plan we have been recom-
mending, the plant is longer growing, stronger in its wood, shorter in
its joints, and more abundant in branches, foliage, and flowers. The
ordinary mode of forcing contemplates no more than removing a plant
from out of doors to in-doors in one year; so that, without having the
advantage of premature ripeness for two seasons, or even one, it has
to perfect its flowers before their time, by great excitement, with a
root hardly established. We hold that a Rose, like a grape vine, cannot, after bearing in the usual season, be changed all at once to early
forcing, without great sacrifice of crop, strength, or beauty. The

fact of sudden excitement being fatal to a Rose is demonstra ed easly enough by the result; take a stong plant, well established, from the cold atmosphere and temperature of the ground, into a full-heated house, and every bloom will be blighted in its incipient state. If a decided change like this is universally fatal, which is the fact, every sudden change, and all approaches to it, are proportionally mischievous. We do not, however, mean to say that roses cannot be forced in a single season, because thousands are so forced and sent to market; and the usual result of such management is, three or four long-drawn branches, with a bud or two at the end of one, and sometimes of two, with scarcely strength to open into a flower. There are exceptions to the choice kinds of roses; in these remarks, we allude only to garden roses. The China kinds are of a different nature, always growing and blooming; winter and summer, if they are kept in a moderate temperature, are almost alike to them, and those which partake of their habit.

The Forcing of Roses—the Dwarf China Kinds.

This family has scarcely any rest in pots, and under protection, it may be merely kept over the winter. There is no place so well adapted for them as a cold pit, with a good dry bottom, and shelves near the glass; but a stout shallow box, with a regular garden light on it, placed high and dry on a paved, slated, or warm, gravelled bottom, makes a good shift.

The China Rose, and all the short-jointed, smooth-barked kinds that are like them in habit, will strike, bud, graft, grow, and bloom any month in the year. The only thing necessary, is to have plants in all stages, and there will never be any want of flowers. In the greenhouse, they continue growing on, and blooming at all times; but they cannot be kept too cool generally, and if abundance of flowers are required on a plant, it must have a previous rest, and be shifted to a warm temperature, and if matted in the roots, a lagre pot, and the heat gradually increased until it will bear that of a moderate stove. All the new growth will flower about the same time, or at least sufficient of it to well decorate the plant. Cuttings may be st uck in the spring, planted out in beds six inches apart, to grow a little; the tops may be pinched off, and the buds taken away all the summer, to make them bushy; and they may be potted up with a compost of half loam, a fourth peat. and a fourth cow dung; trimmed a little into shape, and placed in the shade a while. In September, they may be put into their frames, covered up at night against frost, and opened in mild weather, until the ground freezes; they may then be removed, a few at a time, into an increased temperature, and about a month apart. They will be found to bloom well, and succeed each other admirably, all through the winter and spring, before those out of doors can even fairly start into leaf; the only care required being to syringe them against attack of insects, and if that does not keep them under, fumigate them; and see that they never suffer from want of water. These, however, like the Summer Roses, will force better the second year than the first, by shifting them into pots a size larger, trimming the plants into a proper shape, taking away the weak shoots, letting them rest, and giving but little water towards the end of the summer, except to keep them from actually flagging; putting them in their frames and removing them into heat, as before, a few at a time, and a month apart.

PROPAGATION OF THE ROSE.

THE Rose is propagated by seeds, by cuttings, by layers, by suckers, and by budding or grafting.

Propagation from Seed.

This mode is adopted for the purpose of raising new varieties by crossing different kinds, and is almost exclusively practised by professional florists; it is also employed for obtaining Sweet Briers and stocks. When the seed is gathered in the autumn, it is either rubbed or washed out of the "hips" and kept in dry sand; or the hips are laid in a cool room, and turned over from time to time, till the shell is rotted; the seed is sown in the succeeding spring, after which it will come up the same year.

Sowing of the Seed.—Among the numerous modes of sowing the seed of the Rose, strange as it may seem, the very plan which has been adopted for fifty perennials, or perhaps more, answered as completely as any. For instance, Polyanthus seed and Rose seed were sown in the same kind of soil, loam and dung, in the same sort of pan, placed in the same garden light, watered at the same time; and, though coming up at a different period, submitted to the same treatment in other respects; shaded from the same noon-day sun, and, though at a different time, pricked out into pots, four or five in a pot, round the edge; kept cool, and growing right on; and when the Polyanthuses were placed in their single pots, the Roses were also potted in theirs. They were kept dry rather than otherwise all the ensuing winter, in a cold frame, with their neighbors, well protected against frost; and that was all.

In the spring, when they began to grow, they were bedded out in rows, in a shady border, six inches apart, and the rows a foot apart, and here they remained another season, making considerable growth; some were of the China kind, and those were potted up and kept growing; the others were hooped over with low hoops, which kept the covering close down on them in bad weather, and there were several that died during the winter. · In the spring they were pruned carefully, so far as to remove all but the two or three strongest shoots, and those were cut about half way back. Several bloomed weakly, but most of them made good growth. No part of the success, however, went beyond the growth; not half a dozen came at all double, and though there were some bright colors, there were none in our estimation worth saving. The China ones were rather better, but not good enough; so that, after giving a few of the best another year's chance, every vestige was given or thrown away. The experiments followed up season after season led to the following confirmed practice:—The berries were dried all the winter; they were then bruised in a bag, and the seeds carefully picked out; a slight hot bed was made up as if for annuals; the soil put six inches deep all over, half-rotted turf and half cow dung, raked smooth, and the seed sown evenly and thinly all over—occasionally moistened; the seeds came up well, and were shaded; had plenty of air given, and the usual attendance to see that they were not dry, but not much watered. Here, as soon as they were large enough, they were thinned a little, by carefully removing a few wherever they were too thick, which removed ones were as carefully potted off and kept in the greenhouse. They had no other care during the season than protecting them from too much sun; but they were allowed to be quite open on mild cloudy days, and had warm showers of rain at all opportunities. Here it was found necessary to fumigate them several times to get rid of the aphides, which partially appeared five or six times during the season, but were speedily cleared away. The lights were taken off towards autumn, and the young plants looked as well as could be wished. At the period when frosts were expected, they were removed carefully with all their roots, into a bed made of the same compost, and a foot deep; planted a foot apart every way, and the bed being four feet wide, took four across it, the outer ones being six inches from the edge of the bed. The same precaution was taken with mats and hoops to keep off heavy falls of snow or hard frosts, and they were allowed to push as much as they would, without pruning, all the next season, no other pains being taken than to throw the mat over when the sun was distressingly hot, and to water them freely on dry parching weather, every night. At the autumn, they were replanted, all the weak shoots being cut out, but the strong ones not shortened till spring. Though there was a manifest improvement in the flowers each season, it was four or five before anything like the quality of some present roses was approached.

This practice differs, in some respects, from that of some other nurserymen; we have seen healthy seedlings, since all these pains were taken, where the seeds were sown out of doors in a common bed, raked in like so many onions; came up like so many weeds; grew well and stood the weather without even a shelter from hard frosts. Some may have been killed and not missed, but they did as well, to all appearance, as those more tenderly nursed.

Hastening the Flowering of Seedlings.

When the seedlings come up in May or June, keep them well moistened, but not too wet, until you can get hold of them well to pot off. Put one each into small pots, and let them, as soon as they are established, be placed in the shade out of doors; but the greatest care must be taken to prevent the attack of the fly, or vermin of any kind. They must be looked at almost daily, and upon the least appearance of any insects, you must remove the plants under cover, where you can fumigate and syringe them regularly. It is still better, if you have

frame room, to put them in when potted, because it gives an opportunity of shading, of keeping off too much wet, protecting them
against wind, and of fumigating without the least difficulty, when
necessary. They should, however, seldom have the glasses on.

After the seedlings have been five or six weeks in these pots, they
may be bedded out, in rich beds of loam and dung, without disturbing
the balls; they should be about a foot apart, in beds of four feet wide;
by planting within six inches of the side of the bed, four rows will go
in, and they will here grow rapidly. Before the close of the budding
season, many will have grown quite large enough to bud from; and
the most promising may be cut back, and three or four buds put on
remarkably strong stocks. Select a strong branch for budding on,
and at first, you must let some portion of the branch beyond the bud
be left on to grow; a very small shoot beyond the bud will do to
insure the growth. These buds will strike off vigorously the next
season, and make considerable growth; but before the bud has shot
far, cut the stock away everywhere but the portions budded on. The
growth they will make this summer on strong stocks will insure their
bloom the next season; and, as the real object is to see if the Rose be
good for anything, they should not be pruned, except so far as to cut
away weak branches altogether; by leaving the full length of the
strong shoots, the blooms will be hastened.

In the mean time, those in the bed may be treated as directed; and
though not generally the case under the present management, they
have bloomed these years on their own bottoms, though there were
a great number much later than the third year, and some even went
to the fifth. This mode of budding the promising seedlings hastens
the certainty of bloom very much, as it is very rare indeed that they
miss coming the third year. If they are worth propagating, the
budding greatly increases the quantity of wood to work from. If, on
the contrary, they turn out good for nothing, the instant you discover
it, cut away all the wood, and the stocks will, in all probability, grow
in time for budding other sorts upon the same season you discover
the deficiency of those already worked. In this way, without incur-
ring much trouble, you may satisfy yourself as to the quality of seedlings for a certainty the third year; therefore, you should provide
yourself with stocks for that purpose, whenever you sow seedlings.
For China sorts, you should have some stocks of the common China,

or Boursault, or the Dog Rose, in good-sized pots, and well established; for they may be budded later, protected better, and indeed some of the seedlings which partake much of the China are tender, and really require protection from the frost.

Retarding the Flowering of the Rose.

The most simple method of retarding the flowering of the Provence and Moss Roses, so as to have the plants in bloom late in autumn, is to cut off the tops of the shoots produced in the spring, just before they begin to show their flower buds; the effect of this treatment will be to cause the plants to throw out fresh shoots, which will bloom later, according to the period in which the operation is performed.

It may also be done by transplanting the bushes early in the spring, as soon as they have formed their buds, which should be cut off. The roots must not be allowed to dry before they are put into the earth again; and they will require artificial watering if the season should be dry, to make them flower late in the fall.

Propagation by Cuttings.

When the earliest shoots of the China Rose are about four inches long, cut them off close to the old wood, plant them in pots half filled with soil, and plunge them in a warm situation, placing over the pot a flat piece of glass, to exclude the cold air; the glass should be wiped occasionally. Thus treated, they will make blooming plants by autumn.

Indian Roses, and climbing kinds, are also easily propagated by cuttings and slips, protecting them as above, or by a hand glass, when the climate is cold.

Propagation by Suckers

Many roses, indeed most of them, growing on their own roots, instead of by grafting on a stock, constantly spread at the roots, and branches force their way up, much to the annoyance, sometimes, of the men in charge of the rosary. In the spring months, their suckers should be looked for, and when found, they should be taken off at once, far enough under ground to get a piece of root with them. These should be replanted instantly on the removal; but if a piece be

3

planted out, and devoted to propagation, the proper method is to dig
up the plants in autumn, tracing the roots as far as they go, and taking the portions which have been growing above ground out at the
same time. Some kinds will have half a dozen, or more, perfect plants,
which have been formed by the spreading at the root, and the end
growing up through the surface. These suckers should be trimmed
and planted carefully, at such distance as the sizes warrant; generally
in rows a yard apart, and the plants eighteen inches from each other.
Here they have to be cut down in spring to within three or four eyes
of the ground.

Propagation by Layers.

The Rose will propagate from layers. To do this, some merely select
a lower branch, and, bending the wood sharp between two joints, peg
that down under ground in autumn; it will root well by the following
fall. Others cut a notch in the wood, on the upper side, which makes
the bend sharper; but there is more danger of breaking it. Another
method is, to run a knife through the wood, so as to split it, and then
give the wood a little twist; but most of the sorts will root if only
pegged under the surface. That, however, is rarely resorted to; and
when it is considered what facilities for propagation are offered otherwise, it is no wonder. The laying should be done as soon as the
wood has ripened, and the pegs to be used should be like a miniature
hooked walking stick, which it is easy to form out of any branch of
wood. This hook is thrust into the ground firmly, to hold fast the
whole winter and summer season.

In dry weather, the layers should be watered, as the trees themselves, or bushes, frequently prevent the rain from coming near the
surface, where the branch is pegged down, and they would in such
cases have no encouragement to root. In the autumn of the next year,
examine them all before they are cut off from the parent root, and if
rooted, of which there will be little doubt, cut the new plant away,
with all the new root; and in planting it out in another place, shorten
the portion above ground to half its length; and at pruning time, in
the spring, cut it down within three or four eyes of the ground, in order
that it may form a bush.

Layers of scme roses strike almost immediately ; a. d from this facility, it is a common practice to lay them all over a bed by pegging down the branches on the surface, at small distances, and thus cover a whole space, which have rooted at almost every joint. The flowers, in such cases, are very strong; but a bush thus treated, and every branch layered, would cut up into an immense number of plants.

Propagation by Budding on Briers.

We marry
A gentle scion on the wildest stock,
And make conceive a bark of baser kind
By bud of nobler race ; this is art
Whioh does mend nature—change it rotting ; but
The art i·self is nature.   SHAKSPEARE

There is no process in the art of Practical Gardening more interesting, nor the fruits of which are more gratifying to an amateur, than budding. The theory is this: At the base of the leaf is a small bud, which, after the leaf falls away from it, becomes prominent, and eventually, if left on the tree, makes a branch. By taking a leaf off with part of the bark, this incipient bud comes with it, and by inserting this bark under the bark of another rose tree, say one of these common briers, it unites as if it were originally a part of the brier itself; but the bud retains all the character of the one it came from, and is not changed in the smallest degree by the transfer from its own to another stock. This is the fact upon which all propagation by budding is founded; and, therefore, we l ive two leading points to consider in setting about this operation.

First, we must have the green bark of the stock, into which the buds are to be inserted, rise easily, which it does all the while the branch is green and growing; and, secondly, we must wait until the bud, small and almost imperceptible as it is at the base of the leaf. is old enough to be removed with safety. In a general way, the buds of Summer Roses are not ready till nearly mid-summer, and the bark will not easily rise from the wood of the stock much after that. The budding season may, however, be called from the middle of June to the middle of August, and not very much longer. What is meant by the bark easily rising is, easily leaving the wood, so that it would be easy to peel a branch by stripping the bark off.

The first thing, then, to look to, is to obtain branches of the rose tree from which we want to produce other plants. If you obtain these branches before you are ready to use them, plant the thick end in the ground, and do not let the sun come near them, as it would soon destroy them; but they ought not to be an hour longer than you can help unused. Get some bass matting for ties, or very coarse worsted, which some prefer, because it gives way better if the bud swells, and will stand the weather longer. With a very sharp knife, called a "budding knife," if you have one, and, if not, any other, and a thin piece of hard wood or ivory, like a diminutive paper knife, you may go to work. The knife is to slit the bark down to the wood wherever you mean to put in the bud, and the piece of hard wood or ivory, with a sort of blunt edge like a paper knife, is to divide the bark from the wood by running it along under the bark, on each side of the slit.

Stocks for Budding and Grafting.—The great call for these articles has made it somewhat difficult to procure them anywhere but at the nurseries; and when you consider you can pick and choose at some price or other, the nurseries are the best place for an amateur to purchase. In some parts of the country, the briers are plentiful, but they are mostly in hedge rows, and it is somewhat perilous work to grub them up without permission; nevertheless, many men get their living by collecting these for the nursery grounds. The stocks should be procured at the fall of the leaf, and be straight, strong, well rooted and compact. These should be placed in rows, eighteen inches apart from each other, and three-foot or three-foot-six-inch vacancies between the rows; they should be staked, or, which is better, stakes should be put at equal distances, and a rail along them, to which rail all the stocks should be fastened by strong ties, the whole being well trodden in after the manner that new roses are planted.

The preparation of the roots should be in all respects the same, and the stocks are generally shortened before you get them to the height their growth best adapts them for. Here they remain till they begin to push in spring, when all the lower buds must be rubbed off, leaving the three or four that are highest up the stock to see which will grow best. It will be found that some of these stocks have died down to a considerable distance; but as they are not of the slightest importance above the top growing bud, you may, with a strong knife, cut right down to the bud, or leave it till after the summer growth of the buds has considerably advanced. If you have one good branch, it will do, but two on opposite sides are better, because you can work both, and be safe if one fails. Several times, you must go over these stocks, to rub off the fresh buds that will be springing out on different parts of them, where they are not wanted; and two good buds near the top are all you need save. You have to remember that all the strength of the plant will go into these two branches, if the others are taken rway; but that every leaf that is allowed to grow, besides those wanted, takes greatly from their strength, on which strength the value of the plant entirely depends.

If the top shoots or buds happen to be weak in the first instance, compared with some lower down the stock, it is better to rub off the .op, and lose a little height of the stock, than trust to dwindling branches, so that, in this case, your two branches to save might be half way down the stem; and it is better, in such case, to down at once to it, that the top may be no more trouble, and may not mislead you, in going over them a second time, to cut or pull out your best branches; for the top, so long as you leave it on, would be throwing out its green shoots; and being the same height as the general run of them, nothing is more likely. All that is to be done, besides keeping the stocks from throwing out other branches, is to cut away from the roots any suckers that may come up, and which distress the stock nearly as much as the dwarf branches. The ground, of course, is to be kept clear of weeds until mid-summer, which is the season for budding, and which is the next subject for consideration.

Being thus provided, go to your stocks with your branches of the trees you want to propagate, in your apron; for you ought to have front pockets, and the bass matting should be tucked in the apron

string; take hold of the stock firmly, and shorten both the branches
to a foot, or even less; then with your knife, cut a slit in the bark,
within half an inch of the base of the branch upward, and on the upper
side, an inch and a half long; about the middle of this slit, make a
small-cut across; then with your ivory, or thin wood—or more
properly, if you have it, with the handle of your budding knife—raise
up the bark on both sides; then take the branch of your rose tree

from which you take your buds, and with your sharp knife, shave out
of the branch a thin piece of the wood, beginning half an inch below
a leaf, and taking the knife along to come out half an inch above the
leaf. This small bit has to be inserted under the bark on both sides,
bringing the leaf, which is where the bud is, to the exact place where
the cross cut is; when it is neatly inserted, take your piece of matting

and place the middle of it across the slit just under the leaf; pass it
under, and cross it backward and forward along the branch till the
bark is completely tied down close, and only the leaf and bud exposed

As the weather at this time is often very hot, it is a good plan to tie a bunch of loose moss over all, and water the moss occasionally the first few days, because it keeps off the burning sun, even if dry, and greatly preserves the newly-disturbed bark. It will be easily seen that the quicker this operation is performed the better; because; if the sap of the bud, or that of the raised bark, has time to dry, the union of the one with the other cannot be completed with any degree of certainty.

The bark being damped immediately by the application of wet moss will hardly undo any mischief already done; so that a sharp knife, a clean cut, and rapid action are necessary, and can hardly fail. If the bud is cut out of the branch too thick, and too much wood is taken out with the bark and bud, the wood ought to be cut thinner, or pulled out from the bark of the bud altogether; but there is danger in taking out the wood; for it will occasionally bring out the germ of the bud with it. The effect of this would be, that nothing would indicate outside what was wrong, but the bud would not grow. It would look as green, as fresh, and as completely united, as if the germ were there. On this account, you may omit the practice of taking the little bit of wood from the inside of the bud, and with the greatest success. This operation should be carried through all the stocks, if you have plenty of buds on each of the branches; because two buds will make a head sooner than one, and if you choose to do so, you may put two different sorts on the same stock. In this case, you must be particular about having two of about the same habit; for a fast-growing one would soon deprive a slow-growing one of all the necessary nourishment; and, besides this, it would grow incongruously, and would not be controllable. On the other hand, if you have two of similar habit, and opposite colors, it may be made a very pretty object. But the great value of this delicate, though simple operation, is to make an old China, or other strong-growing Rose, long established, change its face altogether. Many kinds of roses may be budded on such a tree; by selecting all the strong-growing branches of the present year's growth, putting a different bud in each, and cutting all the other parts of the tree away, leaving the novelties alone to grow; or the buds may be all of the same sort, so it be some choice kind; but different colored roses have the best effect.

Spring Budding.—But one of the most sure and expeditious methods is that called "spring budding," by which the bark of the stock, as early in the season as it will separate from the wood, is cut like the letter T inverted, (thus, ⊥) as shown by *a*, in the adjoining figure;

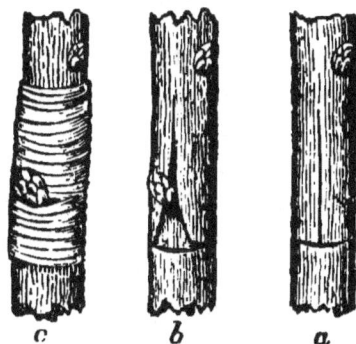

c   b   a

whereas, in "summer budding," it forms a T in its erect position. The horizontal edges of this cut in the stock, and of the "shield bark" containing the bud, should be brought into the most perfect contact, as denoted by *b;* because the union of the bark in spring takes place by means of the ascent of the sap; whereas, in summer budding, it is supposed to be caused by its descent. The parts should then immediately be bound with water-proof bass, (*c*,) without applying either grafting clay or grafting wax. The buds may be inserted either in a healthful branch, or in a stock near the ground. In general, two buds are sufficient for one stock, and these should be of the same variety; as two sorts seldom grow with equal vigor. The bass ligature, which confines the bud, may be removed, if the season be moist, in a month after budding; but if it be hot and dry, not for six weeks at least. As soon as the inserted buds show signs of vegetation, the stock or branch containing them should be pruned down, so as to leave one or two buds or shoots above. If the stock is allowed to have a leading shoot above the inserted buds, and this shoot is not shortened, the buds inserted probably will not show many signs of vegetation for several weeks.

PROPAGATION BY GRAFTING.

This is by means so simple an operation, though not a very difficult matter; nevertheless, the pith in the centre of the wood is against it, as well as the discrepancy in general between the stock and the scion. The act of grafting is adopted for the same purpose as that of budding —to propagate particular varieties. It is not so safe nor so certain a mode as budding, but in the spring, there is no other means; and as in the purchase of new roses, there is generally a good deal of ripe wood that must be cut off, those who have stocks that are fit for grafting frequently adopt it. There are various modes of performing this operation; one or two ways are applicable to the old wood of the stock; other modes are adapted to the last year's branches. In the one case, a cleft is made in the stump of the stock, and the wood belonging to the new Rose to be inserted is cut in an angular form to fit it. It is then bound in its place by bass matting, or some other tie,

and the joins covered with grafting clay, or, which is more generally used for roses, grafting wax; a composition formed of beeswax and resin, in equal parts, and a little tallow, to render it easily fusible at a low heat, because the real object of this wax is to melt at a heat which will not hurt the trees, but that will, on cooling, be sufficiently hard to keep in its place, and bear even the heat of the sun without running away.

There are various modes of grafting the smaller branches of the stock; that is to say, the branches of the last year's growth. One mode is, to cut the branch down to two inches in length, and then cut

3*

this short piece down the middle, cutting out the inside of the wood sloping outward, so as to receive a wedge-shaped graft, which should be about the same size, if possible; cut this into the shape of a wedge, and insert it in the stock, making as complete a fit as possible, and be careful that the bark of both scion and stock exactly join on one side, whether it reach the other side or not; for, unless the barks meet on one side, it will be impossible to unite. It will frequently happen that the scion is smaller than the stock; the one must be used as you have got it, the other you must get as good as you can; and when you have it, make the best of it. Others, in grafting, cut the branch of the stock into a wedge, and the scion is cut to receive it. The effect is the same in the end, if well done, and in good grafting, the joint is soon lost in the growth.

There is one advantage in grafting in spring: If it takes, you may have roses the same year, and thus a season is saved; but, if any of them fail, the stock will grow, if the graft does not; and, of course, if the graft does not grow, you must allow the top branches of the stock to grow, and rub off all other buds, just as if it had not been grafted. The China kinds will graft at any time of the year, but they must be on China stocks, or stocks partaking of the nature of China stocks. It is only the deciduous kind of stock which is confined to the spring grafting, and it is not uncommon to see the solid stock of a large size cleft to make room for a small bit of choice wood; they holding it to

be a waste to throw away the prunings of the Rose, and giving much attention to the profitable use of them.

Root Grafting.—It will be always found in a plantation of roses that suckers spring up in abundance from the roots; these would soon rob the head or worked part of a great portion of its nourishment; but these suckers are useful when taken off with a good portion of root to them, because there is not a more certain mode of propagating the Rose than neatly grafting a piece of the wood of a Rose on the root just under the surface; the union is almost certain, if at all dexterously

done. The proper mode of doing this, is to pull up the sucker, which will expose the root some distance, and take off a good piece of root with it from the parent stock; cut the sucker completely off to the part that was on the surface of the ground; get a piece of the wood of a Rose as nearly the size of the root as possible, cut a slit in the root, making both cuts smooth and flat inside; then cut the scion wedge fashion, and make the bark fit it even with the outer cuticle of the root; tie them well together, and plant them so that the entire graft goes under the surface of the ground. These root grafts are excellent for dwarf plants, for they are worked actually under ground, and when well done they make excellent plants. Grafting the Rose is not chosen before budding; but, as there is always a good deal of waste wood in a rose tree that has to come off in spring, many give grafting a chance; and of grafting, root grafting is one of the most effective. There is never any scarcity of roots among a collection of roses; forking the ground a little brings up these straggling shoots; and so that there be a good piece of healthy stuff, there is no difficulty in making a good job. There is no occasion to clay over the join in · root grafting.

There is another advantage in root grafting: it is applicable with the China kinds all the season through, if you make sure of a healthy root; nor is there any difficulty in obtaining proper roots for the purpose. Wherever a sucker comes up through the ground, use a fork and take up as much root as you think suck a plant ought to have; the operation must be performed quickly, and with a very sharp knife, for the root must not dry under the operation, and they must be planted directly. The graft need not be put in wedge fashion; any other way is as good, if the join be smooth, well fitted, and tied firmly. But we do not recommend grafting of any kind as the best means of propagation. Nothing is so simple as budding, and scarcely anything so efficacious. The propagators of roses by root grafting are very apt to grow the suckers in pots for a considerable time, so that they get completely established after being broken away from the parent root, before they are submitted to the operation of grafting, and this becomes then almost a matter of certainty; whereas we have known the roots of suckers bleed so much, that they have lost the root, and have been indebted to the graft striking root for not losing it altogether.

PRUNING.

THE principal objects to be attained by pruning roses are—first, to compensate, by reducing the part to be nourished, for the loss of the root that has to nourish it, which loss, greater or lesser, is always suffered by removal. The proper way to do this pruning depends much on the state of the plant when you have planted it. If it be very bushy, cut away all the weather branches, leave not more than three or four of the best of the shoots, and shorten even those down to a few eyes. If you wish the plant to continue dwarf and bushy, you may cut down to the last eye or two of the new wood, but leave no thin half-grown shoots on at any rate. If the plant is a matured bush, with numerous branches, and pretty strong generally, shorten the new wood down to two eyes, which will show what more you need do. It may be found that you have then a great many more branches left on than you require; cut one half of them close off, and that half must be the thinnest; but it may be that the plant will be improved by cutting some of the main branches clear away, and all that are on it; for rose trees and bushes, like everything else, are easily spoiled by bearing too much wood, and being over-crowded.

The regular Climbing Rose is often required to make as much show as possible the first year of planting; but unless they are removed with the greatest possible care, they ought to be cut almost to the ground, and thinned out also. None but the strongest wood ought to be allowed to remain on the plant, and if this be not of quite first rate excellence, it is far better to cut out all the weak branches, and cut down the strong ones to two eyes each.

Pruning Standards,

With regard to Standard Roses, we cannot help thinking, from all we have seen practised, that a large portion of them are grown altogether upon a wrong principle. Standard trees, to be handsome, should be as wide in the head as their entire height; and upon the present system of pruning them, they enlarge a little every year.

When your standards are planted, you need do nothing to them until

FIRST YEAR'S GROWTH OF BUD.

April; then cut all small shoots off close; that is to say, clear them right away; cut down the strong ones to two, three, or at most, four eyes. care being taken that the top eye is pointing onward; the object of this is to obtain strong branches growing outward, to make a wide head. As the shoots grow, notice the best and strongest that are growing in a position to widen the head, and leave them to make all the growth they can; allow any shoot that is growing up strong in the centre to grow also; and further, a most important point, rub off, or cut off with a very sharp knife, all weakly growing shoots, all that grow inward and cross the head, and wherever two cross each other, remove the weakest. The branches that grow outward will be good

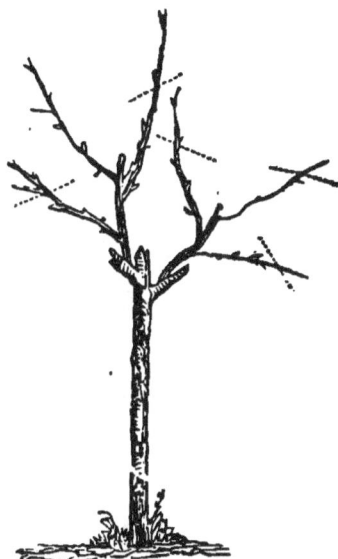

SECOND YEAR's GROWTH OF BUD.

enough and well enough in one season's growth to leave any length you please towards making a proper sized head; but as five or six of these branches will not make a full head, the next season they may be shortened to half their growth, taking care that the end bud must be an under one, for all the tendency of the Rose is to grow upward, and it is only when the natural growth is outward, or downward, that the weight prevails to keep it in a horizontal or drooping position. This second year, and indeed every subsequent year, every branch that does not assist to form a handsome head without crowding, must be taken away, and the younger it is when taken, the more good its removal does, because the other branches get the better.

With regard to any one or two, or even three upright branches, though one strong one is worth three weakly ones, they may be shortened down so that two or three good eyes may be fairly above the other branches, and that when they grow outward the next season, they may help fill up the head of the tree above; when the eyes begin to shoot, rub out all that come where they are not required, and leave those of which you are yet doubtful, as well as those you know will be wanted, because it is at this period you have such control by

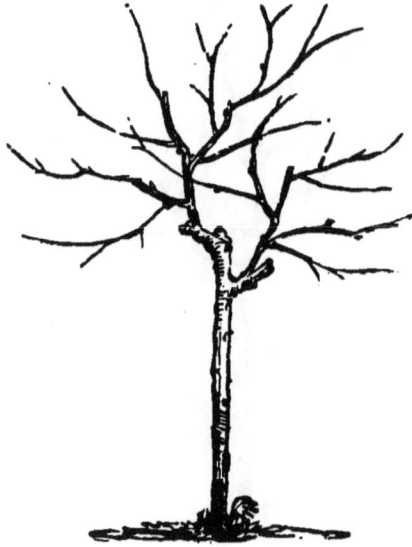

THIRD YEAR'S GROWTH OF BUD.

driving the whole strength of the tree into the branches that are wanted. In this way, you proceed until the head of the tree is the proper form and proportion, instead of, as we now see them everywhere, a small, pimping, ungraceful head to a tall stem, or trunk. When once it has arrived at this perfection, which, with very little care and attention, it will, you may cut back every year's wood to two eyes; cut out every weak shoot altogether, if you have not rubbed it off in the bud; cut out all that are in the way of free growth for the rest, and when any portion is confused by reason of the number of spurs or shortened branches left on, clear away a bit by cutting them off. Always remember that Standard Roses for appearance should not be too closely pruned; but for showing, when the individual blooms are shown, a multiplicity of flowers is against size. We can hardly recommend too strongly the necessity of what we shall call spring pruning, which is, in fact, nipping the mischief in the bud, watching the development of the newly coming branches, and removing all but the number there is good room for; and as this has not been treated of at any length, if at all, we may fairly request attention to it.

The three cuts which are in illustration of this article, though not very accurate, show the first year's growth of two buds placed in a stock, with dotted lines at the place we should cut them; the second year's growth after such cutting, with dotted lines where we should cut them again; and the third year's growth is indicated by lines which give some idea of it. But neither of these cuts is exactly what we like; first, because our pen and ink sketches were imperfect, we being unable to draw exactly what we wanted; and secondly, because the artist, who could have drawn it, did not know what we wanted. There is enough, however, done to assist in our lesson on Pruning Standards, though not to the extent we wished.

Pruning and Training Pillar Roses.

Although we have touched on the pruning of bushes, and upon the pruning of climbers when first planted, it only related to the mere operation of pruning them for growth, in the position they were to remain ; and here, for the sake of the poor roses themselves, and the pillars they are to ornament, we will suppose they are cut down to the ground, or nearly so, and have made a fresh growth, or rather are making fresh growth. Pillars for roses ought to be a foot in diameter, and are best made of trellis work or rods of iron, or, if it must be so, of wood; but they ought to be one foot through. As the leading shoots come, they ought to be wound spirally round the pillar, at such distance from each other as will enable them to fill up the space between with foliage; their leading shoots then constitute the tree, and all the side shoots bear their blooms, and form a pillar of roses. We do no mean that this is all done in a year, though some kinds go a long way towards it; here, as in all other cases of rose pruning, the little weak shoots must be removed, the strongest left on all the way up, and should be shortened to two eyes. If the tops here die down at all, shorten them to the strong top eye, not to the top eye, for several near the top may be found weak, and they would never be otherwise, whereas the stronger one will grow fast, and soon supply the place of the old top.

When the buds first show in spring, it will be right to go over the roses carefully, to remove any that are in the way; and the growth of some roses will be found so different to that of others, that one sort will want enormous room to develop its shoots and blooms, while another will make but short branches and bloom abundantly. These characteristics will be discovered in a year's growth, if not well explained beforehand, and the provision can be made accordingly. Many Pillar or Climbing Roses are made to run over arches from pillar to pillar, or along festoons from pillar to pillar; the best way to manage those parts which form the arch, or festoon, is merely to thin out their weak branches without shortening their strong ones, because they will bloom more abundantly, which is the great charm; and the loose and free manner in which they hang about will be to their advantage, so they be kept within bounds a little.

Pruning and Training Roses on Flat Trellises, Walls, and Fronts of Houses.

The management of this family is very similar to that of Pillar Roses, except that the leading shoots must be encouraged to grow the best way to fill up the space allotted to the plant, for which purpose it will be advisable, in some cases, to train the strongest two shoots horizontally right and left along the bottom; or if the space to cover be only one way, to train one strong shoot along the bottom, and turn it up at the end; if it reach further, the rest of the strong shoots may be fanned out at equidistances, and all the weak joints removed. The next year, rub off the buds that are coming where they are not wanted. Allow any strong shoots that come up from the bottom horizontal shoot, to grow as much as they will, but no weak ones. A fast-growing Rose will soon cover a house front, a trellis, or wall, and flower all over.

When the space gets filled, you must continue cutting out, from year to year, all thin, spindley shoots, and spare the strong ones, so that the stongest eyes only are developed, instead of all of them; and the Roses are closely set to their wall or trellis, instead of hanging lolloping about; the very thing which is good on a pillar, or an arbor, or over an archway, or on festoons, being the reverse on a flat surface. As a never-failing operation, however, in all cases, the weak, spindley shoots may always be removed, whether the strong wood be shortened or not.

Pruning of Standards on Their Own Bottoms, or Roots.

It is very common to see among Dwarf or Bush Roses, a strong shoot growing upright, a sucker from the root; and it is frequently the case that these will rise up to five or six feet high. In the Moss Roses, this is often to be found. These may always be trained into standard trees, with heads in every way proportioned to the stem. As soon as a vigorous shoot of this kind makes its appearance, cut in the bush at bottom rather hard, as it will tend to strengthen the root, which will be relieved of some of its work by the operation. When the shoot has attained the required height, pinch off the top; this will encourage side shoots, all of which, except the two or three at the top, must be rubbed off. It rarely occurs, however, that any side growth is made the first season; so that the better way, unless the shoot be getting too long early in the season, is to let it ripen its wood. The latter part of the autumn, you may look at the root, to see what state it is in, and how far it may be dependent on the main root. If it be closely joined, so that there would not be sufficient root if separated, the old bush must be sacrificed, and the root secured for the standard. As the upper part of the shoot may not be well ripened, it will be as well to bind a hay band round it, or tie some moss or other litter, to save it from sharp frost, though moderate ones will not injure.

In the spring, cut the end off as low down as will do for your purpose, and when the buds shoot out, it will be seen that the three or four upper ones come first; all others on the stem must be rubbed off. Nor is it any great use having two buds on the same side of the tree; f you can manage to have three, or even four, within a few inches of the top, pointing different ways, they will form the better hold of it, to strengthen the other portion of the tree. Continue to be watchful as to other buds that will be continually pushing from the main stem, and let not one grow but those you have selected for the head.

At the end of the year, these will have made considerable growth, and, instead of being cut back the next spring to two eyes, as is the case with many, cut them back only so far as to insure the strength of the remainder, say, so as to leave five or six eyes. The next season of growth, there will, out of three or four branches, come four or five branches each. Those which come in their places, to help form a handsome head, may be allowed to grow; but if any come so as to cross others, or where there is plenty of growth already, let them be rubbed off; but it is quite possible for an eye to shoot where it is not wanted, and yet the first or second eye of that shoot may be in a direction to fill up a vacancy where it is necessary; this must, of course, be looked to before buds are rubbed off. These branches, when grown another season, will stretch out the head on all sides to a respectable size, and enable you to thin out the weak wood, and cut back the strong; so that instead of having the head pimping and small, it may bear a proportion to the stem; for, as we have said before, the head ought to be as wide across as the stem is long from the ground, to the under part of the head. There is one thing to be observed with regard to standards on their own bottom: they never break off, nor decay, nor canker, half so much as budded and grafted ones.

GENERAL HINTS.

WE may mention, as a general characteristic, that there is no plant which yields more willingly to culture than the Rose, nor in the growth of which there is so much certainty. If you desire a large quantity of bloom, and are not anxious about the size of the flowers, there is nothing required but to spare the knife; take out weak shoots, but leave plenty of wood on the tree; for every eye will bloom, and

the more you leave on, the better for that purpose. In this case, the
new wood made is but short, because there is so much of it. If, on
the contrary, you desire large blooms, cut away all the strong wood,
of the year previous, down to two eyes at the most, and cut all the
weak wood out altogether. Indeed, you may go further; for you
may cut away half the strong shoots, and lessen the number of eyes
still more.

Again, roses in poor soil will grow and bloom; their flowers will be
smaller, but not less healthy; their wood will be weaker and shorter,
but still sound. The principal danger when a Rose is starved is, that
it may come less double; and this is so serious a fault, that it has
occasioned many to be thrown away that did not deserve it, and
caused many others to be considered wrong varieties, when they
wanted nothing but good growth to make them right ones. On the
other hand, rich soils will cause a Rose to grow enormously; and all
intermediate growths between the strongest and the weakest may be
secured according to the soil they are put in to grow. Generally,
people fancy that dung is the only thing required; this is a mistake,
loam is required to grow the Rose in perfection; and if the ground is
poor and light, a spadeful of loam and a spadeful of dung will be far
better than two spadefuls of dung. This ought to be always mixed
with the soil a little, and the Rose planted in it.

Roses are sadly injured by the wind, and the blooms require fasten-
ing to something or other, to prevent their being frayed. The stakes
of roses should always be made fast to the Rose, or the roses made fast
to the stakes with leaden or copper wire; because bass matting, or
other perishable stuff, will give way when high wind takes them, and
they receive a good deal of mischief before they are observed and
fastened again.

Of the roses at present in cultivation, very few which are not semi-
double will open out boldly; and those which are semi-double, are
not fit to show as single flowers. There are, however, some which
will bear the test of stand-showing, and they not of the dearest or
newest. Those, therefore, who desire to grow none but perfect
flowers, should state to the dealer, of whom they mean to buy, that
their object is to have none but such as will expand and show a good
face when fully bloomed, as they purpose growing none others. The

establishment of the showing in stands, like dahlias, will cause many old and fine roses to be appreciated, and a great many new ones to be discarded; for although it is not the gayest mode of exhibiting roses, it is by far the best mode of testing, and it is curious to see the number of varieties with very glaring faults. For instance, some are close balls of petals, with the outer ones rolling back a little, as if they were shrivelling; but never opening fairly. Others no sooner open than they show their yellow seeds and their paucity of petals; some are on stems too weak to hold them in their position; others, again, burst into a broken mass of ill-formed petals, that do not compensate for their sweetness. Some fall to pieces the instant they are open, and others almost before they open; many are shapeless masses of colored flimsy texture, that neither hold themselves in form nor impart fragrance. It is worth while to direct the attention of the amateur to the large collections of roses sometimes to be seen at exhibitions, and to the very few which are to be found among them of a fine form. They will observe bunches of half-bloomed flowers, that dare not be shown; they will find plenty of hard lumps, on stems not strong enough to bear them without lolloping about; they will find some without a round smooth petal among them, but very few so good as the Tuscan, the Cabbage, the Moss, the Provence, and the oldest of the known good varieties. This shows the necessity of attention to the hints we have thrown out; for we must again confess, that although we have selected the best among eleven or twelve hundred roses, there are many that we shall see rejected like the remainder of the entire collection, to make way for better flowers and better taste.

As a concluding observation respecting the management of the Rose, we are bound to say, that a good deal that is done now is erroneous, although taught by rose cultivators; and especially with regard to roses in pots, which, however pretty they may look, are very much drawn, and very unnaturally supported. That the system, if pursued, will lead to the introduction and toleration of varieties which cannot support themselves, in the same manner as it did to the introduction of worthless geraniums, there is no room to doubt; for in the specimens exhibited in pots at various shows, the total inability of the flowers and stems to support themselves is manifested, as well as the dispositions to encourage this strange mode of distorting things. Some allowance should be made for any forced subject; but that gardener who can produce his plants without supports, is he one who deserves a prize for his skill; not the man who draws a plant till it cannot support itself, and then keeps it up with framework.

There is much to be done in the choice of roses, for particular objects. Those inclined to droop should be on very tall stalks, for their pendulous habit is very handsome, and renders the tree a beautiful drooping object; those for bushes ought to be short jointed and close habited, as best suited to dwarfs, and so also will they be found for dwarf standards.

The general routine for rose culture is given both as respects the general collection, and also for seedlings; and with attention to what has been here written, we think a mere novice may, with a little enterprise, beat one who grows upon any other system.

Few people are aware of the injustice sometimes done to roses, which are condemned as worthless, when the culture alone is the cause of their misbehavior. The Rose is a fidgety customer. The French people are famous for raising new varieties, and describing them as very superb; the English and American nurserymen buy them as soon as they can be obtained, and describe them to their customers as something *recherché;* they are purchased by amateur cultivators upon the strength of such characters, grown for a year, and too often thrown away as worthless. Once for all, let us inform our readers, that no Rose can be depended on for growing to its character under the third season. The effect of poor culture is to make a Double Rose semi-double and single; and that which would be rich culture to anything else, may be poor to the Rose, because if it be not suitable, it may as well be poor.

There are many things which affect the Rose, but the principal one is tantamount to saying that it does not feel itself at home. European nurserymen often propagate roses rather too mechanically ; the greater part of them are "made to sell." So long as the stock will keep the bud alive, and let it grow, that is all the nurseryman asks or wishes. Now, it is quite certain that a stock without much root will live, and hundreds of plants sent from abroad are of this description. There may be strength enough in the stock to grow and bloom the kind upon it, but as the stock is not fairly at home, the first year is often wasted in making root enough to lay hold of the ground, and during this period, the head is grown but poorly.

As to blooming, it should not be allowed until the growth is vigorous, for it comes miserably poor, if at all. The second year, it is more reconciled to its place, and the third may be considered a fair trial. Take the very best Rose we have, and grow it badly, the result will be bad flowers; but, if this be the case with well-known varieties, how cautious ought we to be of condemning a candidate for our favors when we have no evidence of its real character. Rose growers say it is impossible to tell, after a removal, what a Rose ought to be by what it is; that it ought to be tried three seasons before condemnation, and not be discarded under an idea that it is useless, merely because it flowers badly, which is not always the case. A Rose will sometimes be for several years only middling, when, if it liked the ground, it would be excellent.

When you have a Rose, first you should cut away all bruised parts of the root, and see that all the broken ends of the shoots in the ground, or root shoots, are smooth; then plant it the first year in good strong fresh loam, from a pasture. If rotten dung be at the bottom, so much the better, but do not let the dung touch the roots. Cut nothing back of the head or bud shoot, or if it be an established head, cut nothing back until you see the buds swelling, so as to enable you to calculate what portion is alive, and what has died back. As soon as this is indicated by the growing of the buds, cut away clean to the tree all the branches which may have perished. When these are removed, you see what head you have to depend on, and how much you may cut back without losing an opportunity of forming or improving a head for the next season. For instance, all the branches but one will often die back, and be forced to be removed by the knife. Had the pruning at first been close, and each branch cut back to two eyes, there would be but two, of course, left on the only living one, and but two shoots could be had from them; having, however, discovered that but one branch is left, this has to be preserved somewhat longer, and therefore should be pruned to four or five, instead of two eyes. These nay be managed to form branches all round the tree, or rather at such distance as prudence dictates, due regard being had to the strength of the plant. If the tree takes off vigorously, and the wood grows very strong, the bloom is pretty sure to be inferior, as indeed is the case when almost any plant runs to wood; so that it is quite as unlikely that the bloom of the Rose is in character when the plant is too vigorous, as when it is meagre or too much starved

Hybridising has done much good for roses, but it has also done its mischief; for, if it has introduced some splendid varieties, it has teased us with hundreds not worth growing; some, which are close hard lumps of rolled-up petals, turn over their thin edges like a dog-eared book; the backs of the petals a dull color, scarcely any scent to them, and altogether bad openers, and bad if they can be made to open. There is no reason why the Rose should not be as perfect as the Camellia japonica. There are some of the Bourbons with petals as smooth and as thick, and almost as regular; and these are the kinds to buy and grow. They hold their form longer and better than those with thin petals; they open more freely, and are better when they do open. The habits of these full-flowered plants are better; the flowers, instead of lolloping their heads down, show themselves well. All the full free opening roses of old age are of this description; witness the Cabbage Rose, the Maiden's Blush, the Provence, and some others, which are as familiar as the name of the Rose itself. It is true that the bud of a Rose is pretty, and that a bunch of roses is pretty, but while we have good roses that will open, and of almost every color, it is unnecessary to grow bad ones; and if the character of roses is established by showing single blooms, which shall be required to be open, there will be but little difficulty in doing all the rest.

Nevertheless, on receiving roses from nurseries, whether American or foreign, pay attention to these directions in the planting, and be not in a hurry to condemn. Let them fail the first season, and be even middling only the second, but give them the benefit of the doubt, and try them a third season. Convince yourself that the variety is incapable of becoming better, and that you have seen their natural habit, before you throw them away. If a petal is thin and curly, rough-edged and flimsy, it can never be good; if the petals are good, but there are too few of them, there is great hope that culture will improve it from a semi-double to a perfect double, which is all that is wanted.

CALENDAR OF OPERATIONS.

THE following Calendar for the management of the Rose, during each month of the year, is designed for the central parts of the United States, including the temperate regions of Maryland, Virginia, Ohio, Kentucky, Indiana, Missouri, and Illinois. The season of spring commences in the middle latitudes of Georgia, Alabama, Mississippi, and of Texas, and the northern part of Louisiana, and the southern part of Arkansas, about one month earlier; and a month or five weeks later in Rhode Island, Massachusetts, and in the central latitudes of New York, Wisconsin, and of Michigan. The period of sowing, however, will admit of some latitude, on account of the degree of dryness of the soil, and of its exposure to cold or moist winds, and to the solar warmth.

It has long been observed that Nature, in her operations, is so uniform, that the forwardness of trees, in unfolding their flowers and leaves, is an unerring indication of the forwardness of spring; and that the period at which the shrub red bud (*Cercis canadensis*) puts forth, is the proper time to plant Indian corn, and sow in open culture the seeds of the Rose.

January.

Look well to all standard roses; see that their stakes are firmly in the ground, and the stocks or trunks are well fastened to them. If the heads of standards are very large, compared with the hold they have upon the stock, it is necessary that the stock to which the tree is fastened should reach partly through the head, and be fastened to the head itself. It is also desirable, when very large growth has been made, to shorten, though not properly prune, all the longest branches, to lessen the head, that the wind may not have too much power. If you have not provided yourself with stocks before this month, lose no time, and when procured, prune the roots into moderate form, for they will frequently be found straggling and awkward. Besides planting out a number in rows, to be worked in the open ground, pot some of the most compact-rooted in pots, and plunge them, making a post-and-rail sort of frame along them to fasten the stocks to, and prevent them from being disturbed by the wind; also, if you have not got in all the roses you want, order them and plant.

Protect the smooth-wooded kinds, budded on the stocks, in pots, from the cold, and see that those in beds are well covered with litter where there is danger of their suffering from frost; and, as the smooth-wooded varieties budded in pots will be growing, support their shoots and remove all other eyes from the stocks the instant they break.

At the *North*, where roses in parlors and greenhouses are coming into flower, syringe the plants freely with water, and occasionally with a solution of Peruvian guano, mixed in the proportion of half a pint of guano to eight gallons of water. Fumigate often with tobacco, in order to keep down the green fly; and with sulphur, to kill the red spider.

February.

Look over the established stocks, and see which are most favorable for grafting; and if you have any wood of roses you intend to graft, leave it on the trees; but if you have to obtain wood, seek for it in time; and if you get it, plant the thickest end downward in the ground, in some shady place, because they ought not to be grafted till next month, and the cuttings will keep some time. The China Roses in the house, and roses in the forcing house, must be kept well syringed, and watched carefully, that, in the event of the green fly attacking them, they may be fumigated, as well as syringed. Roses in pots should be kept a little moist, and if not pruned in autumn, should be pruned directly. Look to a supply of wild stocks, if you have not yet com-pleted your arrangements.

Bruise the berries which have been saved for seed, and rub out the seed ready for sowing next month.

At the *North*, continue the same treatment as recommended last month.

March.

Prune all roses which were left half done in the autumn, or not done at all, especially grafted and budded ones of last year, as they have this year to make some growth. Stocks may still do if the season is backward but not a day must be lost. Look over all the standard trees, examine the pushing buds, trim out all weak shoots from the buds, and cut away all shoots from the stock. This must be always considered of first consequence, for the growth of a branch from a stock will completely check the growth of the head. All grafted and budded trees, when once fairly growing, should be deprived of all means of growth from the stock itself. It is not wise to destroy altogether the growth of the stock above the graft or bud, until the union and sub-sequent growth of the graft or bud itself are well established; but this once accomplished, leave no vestige of growth belonging to the stock, and constantly rub off every bud. You may commence grafting this month.

Sow the seeds in large pots or deep pans, and keep them from once getting dry, or being frosted.

At the *North*, hardy roses may be safely pruned the last of this month.

April.

If any suckers appear among established roses or stocks, worked or unworked, remove the earth down to where they join the root, and cut them off close. . If the rose quarter is at all infested with snails or slugs, use all means to destroy them. Inverted flower pots, tilted on one side, will catch many snails; cabbage leaves laid on the ground, and examined daily, will entrap slugs. All stocks on which grafts or buds failed last season must be looked upon as new stocks, and cut down to where they appear alive. The shoots upon which buds were placed should be cut off close, as well as side growths, if they are intended for budding, but if for grafting, the inside shoots may be strong enough to graft on; if the grafts, and the shoot grafted on, be nearly alike, the graft may be all the safer, and the place of union more completely healed than when small grafts are placed in large stocks. This month is a good one for grafting or spring budding, though the operation may be performed successfully in March. Cut back to two eyes all that have been left unpruned, by which late pruning back, the blooming will be protracted considerably.

Keep the seeds sown last month, moist; and if the season be dry, moisten them by laying on the surface some wet moss. Shade them, also, from the hot sun.

At the *North*, hardy roses of all kinds should now be pruned, Moss

Roses cut back short. Rose seeds may be sown the last of this month or early in May. Spring budding may also be performed.

May.

This is an important month with the Rose. First and foremost, the vigilance in looking for the breaking buds of stocks, which would rob the head of its growth, must be doubled, and every three or four days they must be examined and rubbed off. Suckers must also be grubbed up the instant they appear. The shoots of the buds of last year will make rapid growth, and require to be screened, that the wind may not break them out or damage them; and it is a very good plan to tie a stick to the stem, to reach a foot above it, and this does well to support any of the shoots. But when a bud throws up a very strong shoot, it is well to take the top off as soon as there are two pair of leaves, for it will make the shoot form a head the first season; but, in any case, the shoots must be supported by a loose tie to the stick above mentioned.

The young seedlings will be up this month, and will require great care to keep them from damaging by too much wet, or burning up for want of moisture.

At the *North*, Tea, Bengal, Noisette and other roses may now be planted out in borders. Rose seed may be sown early in this month, and spring budding performed.

June.

This month, great diligence must be used to prevent the stocks from growing from their own wood, instead of throwing all their stength into the grafts and buds. It is time also to be looking out for sorts you intend to bud with, either by buying the plants outright, or bespeaking buds for the season; and if any come in your way about the end of the month, do not be afraid of budding on the strongest wood you can find of the present season's growth among the stocks, though you may properly choose a later season, if you have nothing to hurry you.

The young seedlings will have advanced enough to pot off, one in a pot, with loam, peat, and decomposed dung; they must be placed in the shade out of doors, or in a frame and light, in order to grow five or six weeks. See that they are watered as often as may be necessary; and on any appearance of the green fly, fumigate them with tobacco.

At the *North*, roses should be planted out in borders for summer blooming.

July.

If this month be at all forward, you may bud; and if you have wood given to you when you are not ready for it, put the ends in wet sand, and a hand glass over them; but the sooner you can use the buds after you have got them the better. The stocks must be put in completely all over, except one or two eyes beyond the bud on the branches in which the bud is inserted. All China Roses in pots or out of doors may be budded, and so also may all the smooth-barked kinds.

Plant out the young seedlings potted last month, in beds four feet wide, in the same soil, without disturbing the balls of earth; let them be six inches from the side of the bed, and a foot apart each way. Protect them from vermin by all ordinary means; shade them from the heat of the sun at mid-day; water if required.

At the *North*, roses of all kinds planted in open ground, may be layered the last of this month. Perpetual Roses will bloom best in autumn, if they are pruned in after having opened their first flowers.

August.

Continue the budding, and use every precaution to prevent the stock from growing, and remove suckers the instant they appear above ground. Nothing should be allowed to grow, except, just beyond the bud; a shoot may be beneficial, as it draws the sap past the bud; but as soon as it is united and doing well, anything growing beyond it may be broken off, or bent down to check it a little. Cuttings of the smooth-barked kinds will strike almost every month in the year; but at the end of this month, whatever you may be anxious to propagate may be struck in the shade, under a hand glass, or even quicker where there is a little bottom heat.

The same directions will also apply to the *North*.

September.

You may n₁ w examine the budded plants, and undo the ties of any that appear to swell, tying them more loosely, although tight enough to hold in the bud. If any of the buds have failed, you may open a fresh place, and insert others; but if well done, this will seldom be the case. Continue to remove any shoots or eyes that are showing growth in the stock, for on this much depends. Cuttings from the smooth-wooded kinds may be taken and struck, and any that are struck may be potted off in small pots. Weed the young plants in the beds. Water them if the season be dry.

At the *North*, roses intended for early forcing, should now be repotted and pruned.

October.

Towards the end of the month, look out for healthy stocks, or get some one in that way of business to collect for you. Always choose, and make any one who undertakes to supply you understand that you require strong stems, perfectly straight, with compact roots, that have not been much damaged by removal. Any that you get should be at once trimmed and planted in rows, about eighteen inches apart, and the rows wide enough to enable you to go up and down them well, to operate in the way of grafting and budding when required. Many of the budded stocks may now be untied altogether, but it is not well to cut the branches in which they are budded close down to the bud until the spring months. As they would be more susceptible of damage by frost, let them all be properly sheltered, and fastened, if they have become loosened. Shorten the longest branches of standard roses, that they may not hold the wind so much; and although it would be improper to prune, their close back branches may be cut clean away, because they are of no use on the tree. Cuttings of the China and smooth-wooded kinds may be taken now for general propagation. The plants will be the better for losing the wood, especially all the dwarfs in pots.

Examine the August-budded plants, and loosen the ties, if necessary. Break or cut off the wild part of the stock above the bud, all except one growing eye, to keep up the circulation; remove all other branches and shoots. Gather the "hips," or berries, of any desirable varieties for seed, as soon as ripe. Look to those roses budded on stocks in pots.

At the *North*, all tender kinds, growing in open ground, should be taken up and potted, and hardy roses may be successfully transplanted the last of this month.

Nobember.

This is the best month in the year, if the weather is dry and open, for planting out the garden sorts of rose trees and bushes; therefore, all removals should be performed as soon as convenient, according to the plan pointed out in the foregoing treatise. The leaves of all the garden sorts are falling, or have fallen. Some of the perpetuals, and the China and hybrid kinds, are, in mild autumns, still growing, and perhaps blooming. Such must not be touched till the leaves have turned yellow, or have dropped; but in all other cases, where the leaves have faded, the removal is kindly and beneficially done. Stocks may be procured and planted, and if the permanent planting cannot, for any particular reasons, be done now, they must be temporarily planted or laid in the earth, in a sloping direction, and the roots well covered with mould, which must also be well shook in among the roots and fibres. Cuttings may still be made of the smooth-wooded kinds, and placed close together in pots of mould, with half an inch thickness of sand at the top. These pots must not be allowed to dry, but may be put in a pit or greenhouse, or plunged under a hand glass in the border, which will answer for covering them well from frost.

In all situations subject to frosts, throw light litter, as pea vines, pine boughs, or straw, over the beds containing tender varieties, at night; and if there happen to be frost, do not remove the litter during the day. Continue to gather ripe berries, or hips, as directed last month. Cut out the weak shoots from the seedlings, leaving only the robust and strong ones on the plant, except such as are intended for buds in the spring.

At the *North*, tender roses should all be taken up this month. Perpetuals and Bourbons, in the open ground, if in a well-drained situation, with a little covering, will stand the winter without injury.

December.

Planting goes on well this month, if the weather be dry and open; but if wet, and the ground does not work well, it is better deferred; for if planting is done when the soil will not crumble well, and go between the roots, they cannot succeed. Look well to last month's directions, and attend to them in all respects, if not done before.

· Seed berries, designed for sowing next spring, may be preserved by putting a tile at the bottom of a flower pot, into which may be put those hips that are perfectly ripe, covering them three or four inches with sand, and let them remain until wanted; or lay them on a shelf to dry out the moisture. See, also, that the stocks, which have been budded, are secured to stakes against the effects of the wind. Protect the smooth-wooded kinds, budded on the stocks in pots, from the frost, and look well to the litter on those in beds.

At the *North*, those roses, taken up and potted last month, should now be headed in, cutting away all small shoots to one good eye. They may be wintered in a cold frame, or taken into the house, where they will bloom from February to May.

INSECTS.

THE insects which infest the Rose are quite numerous; but as their habits are comparatively but little known, it has thus far been very difficult to arrest their ravages, or sensibly diminish their number, by artificial means. At least forty distinct species are described by European naturalists, but many of them do not exist among us. The only reliable authority on this subject, in this country, is Dr. T. W. Harris, of Harvard University. From his "Report on the Insects Injurious to Vegetation in Massachusetts," we copy the following, which, doubtless, will be acceptable to all who are not in possession of his work:—

The saw fly of the Rose, which, as it does not seem to have been described before, may be called *Selandria rosæ*, from its favorite plant, so nearly resembles the slug-worm saw fly as not to be distinguished therefrom except by a practised observer. It is also very much like *Selandria barda, vitis* and *pygmœa*, but has not the red thorax of these three closely-allied species. It is of a deep and shining black color. The first two pairs of legs are brownish grey or dirty white, except the thighs, which are almost entirely black. The hind legs are black, with whitish knees. The wings are smoky and transparent, with dark-brown veins, and a brown spot near the middle of the edge of the first pair. The body of the male is a little more than three twentieths of an inch long, that of the female one fifth of an inch or more, and the wings expand nearly or quite two fifths of an inch. These saw flies come out of the ground, at various times, between the twentieth of May and the middle of June, during which period they pair and lay their eggs. The females do not fly much, and may be seen, during most of the day, resting on the leaves; and, when touched, they draw up their legs, and fall to the ground. The males are more active, fly from one rose bush to another, and hover around their sluggish partners. The latter, when about to lay their eggs, turn a little on one side, unsheath their saws, and thrust them obliquely into the skin of the leaf, depositing in each incision thus made a single egg. The young begin to hatch in ten days or a fortnight after the eggs are laid. They may sometimes be found on the leaves as early as the first of June, but do not usually appear in considerable numbers till the twentieth of the same month.

How long they are in coming to maturity, I have not particularly observed; but the period of their existence in the caterpillar state probably does not exceed three weeks. They somewhat resemble the young of the saw fly in form, but are not quite so convex. They have a small, round, yellowish head, with a black dot on each side of it, and are provided with twenty-two short legs. The body is green above, paler at the sides, and yellowish beneath; and it is soft, and almost transparent like jelly. The skin of the back is transversely wrinkled, and covered with minute elevated points; and there are two small, triple-pointed warts on the edge of the first ring, immediately behind the head. These gelatinous and sluggish creatures eat the upper surface of the leaf in large irregular patches, leaving the veins and the skin beneath untouched: and they are sometimes so thick that not a leaf on the bushes is spared by them, and the whole foliage looks as if it had been scorched by fire, and drops off soon

4*

afterward. They cast their skins several times, leaving them extended and fastened on the leaves; after the last moulting, they lose their semi-transparent and greenish color, and acquire an opaque yellowish hue. They then leave the rose bushes, some of them slowly creeping down the stem, and others rolling up and dropping off, especially when the bushes are shaken by the wind. Having reached the ground, they burrow to the depth of an inch or more in the earth, where each one makes for itself a small oval cell, of grains of earth, cemented with a little gummy silk. Having finished their transformations, and turned to flies, within their cells, they come out of the ground early in August, and lay their eggs for a second brood of young. These, in turn, perform their appointed work of destruction in the autumn; they then go into the ground, make their earthen cells, remain therein throughout the winter, and appear in the winged form, in the following spring and summer.

During several years past, these pernicious vermin have infested the rose bushes in the vicinity of Boston, and have proved so injurious to them, as to have excited the attention of the Massachusetts Horticultural Society, by whom a premium of one hundred dollars, for the most successful mode of destroying these insects, was offered in the summer of 1840. About ten years ago, I observed them in gardens, in Cambridge, and then made myself acquainted with their transformations. At that time, they had not reached Milton, my former place of residence, and have appeared in that place only within two or three years. They now seem to be gradually extending in all directions, and an effectual method for preserving our roses from their attacks has become very desirable to all persons who set any value on this beautiful ornament of our gardens and shrubberies. Showering or syringing the bushes with a liquor, made by mixing with water the juice expressed from tobacco by tobacconists, has been recommended; but some caution is necessary in making this mixture of a proper strength, for if too strong, it is injurious to plants; and the experiment does not seem, as yet, to have been conducted with sufficient care to insure safety and success.

Dusting lime over the plants, when wet with dew, has been tried and found of some use; but this and all other remedies will probably yield in efficacy to Mr. Haggerston's mixture of whale-oil soap and water, in the proportion of two pounds of the soap to fifteen gallons of water. Particular directions, drawn up by Mr. Haggerston himself, for the preparation and use of this simple and cheap application, may be found in the "Boston Courier," for the twenty-fifth of June, 1841, and also in mos of our agricultural and horticultural journals of the same time. The utility of this mixture has already been repeatedly mentioned in my treatise, and it may be applied in other cases with advantage. Mr. Haggerston finds that it effectually destroys many kinds of insects; and he particularly mentions plant lice of various kinds, red spiders, canker worms, and a little jumping insect which has lately been found quite as hurtful to rose bushes as the slugs or young of the saw fly. The little insect alluded to has been mistaken for a species of thrips, or vine fretter; it is, however, a leaf hopper, or species of *Tettigonia*, much smaller than the leaf hopper of the grape vine, (*Tettigonia vitis*,) and, like the leaf hopper of the bean, entirely of a pale-green color.

In treating of the common Rose Bug, or Rose Chafer, (*Melolontha subspinosa*,) Dr. Harris says:—

The natural history of the rose chafer, one of the greatest scourges with which our gardens and nurseries have been afflicted, was for a long time involved in mystery, but is at last fully cleared up. The prevalence of this insect on the Rose, and its annual appearance coinciding with the blossoming of that flower, have gained for it the popular name by which it is here known. For some time after they were first noticed, rose bugs appeared to be confined to their favorite, the blossoms of the rose; but within thirty years, they have prodigiously increased in number, have attacked at random various kinds of plants in swarms, and have become notorious for their extensive and deplorable ravages. The grape vine in particular, the cherry, plum and apple trees, have annually suffered by their depredations; many other fruit trees and shrubs, garden vegetables and corn, and even the trees of the forest and the grass of the fields, have been laid under contribution by these indiscriminate feeders, by which leaves, flowers, and fruits are alike consumed. •

The unexpected arrival of these insects in swarms, at their first coming, and their sudden disappearance, at the close of their career, are remarkable facts in their history. They come forth from the ground during the second week in June, or about the t.me of tne blossoming of the Damask Rose, and remain from thirty to forty days., At the end of this period, the males become exhausted, fall to the ground, and perish, while the females enter the earth, lay their eggs, return to the surface, and, after lingering a few days, die also. The eggs laid by each female are about thirty in number, and are deposited from one to four inches beneath the surface of the soil; they are nearly globular, whitish, and about one thirtieth of an inch in diameter, and are hatched twenty days after they are laid. The young larvæ begin to feed on such tender roots as are within their reach. Like other grubs of the Scarabæians, when not eating, they lie upon the side, with the body curved so that the head and tail are nearly in contact; they move with difficulty on a level surface, and are continually falling over on one side or the other. They attain their full size in autumn, being then nearly three quarters of an inch long, and about an eighth of an inch in diameter. They are of a yellowish-white color, with a tinge of blue towards the hinder extremity, which is thick and obtuse or rounded; a few short hairs are scattered on the surface of the body; there are six short legs, namely, a pair to each of the first three rings behind the head; and the latter is covered with a horny shell of a pale rust color. In October, they, descend below the reach of frost, and pass the winter in a torpid state. In the spring, they approach toward the surface, and each one forms for itself a little cell of an oval shape, by turning round a great many times, so as to compress the earth and render the inside of the cavity hard and smooth. Within this cell, the grub is transformed to a pupa, during the month of May, by casting off its skin, which is pushed downward in folds from the head to the tail. The pupa has somewhat the form of the perfected beetle; but it is of a yellowish-white color, and its short stump-like wings, its antennæ, and legs are folded upon the breast, and its whole body is inclosed in a thin film, that wraps each part separately. During the month of June, this filmy skin is rent, the included beetle withdraws from its body and its limbs, bursts open its earthen cell, and digs its way to the surface of the ground. Thus the various changes, from the egg to the full development of the perfected beetle, are completed within the space of one year.

Such being the metamorphoses and habits of these insects, it is evident that we cannot attack them in the egg, the grub, nor the pupa state; the enemy, in these stages, is beyond our reach, and is subject to the control only of the natural but unknown means appointed by the Author of Nature to keep the insect tribes in check. When they have issued from their subterranean retreats, and have congregated upon our vines, trees, and other vegetable productions, in the complete enjoyment of their propensities, we must unite our efforts to seize and crush the invaders. They must indeed be crushed, scalded, or burned, to deprive them of life ; for they are not affected by any of the applications usually found destructive to other insects. Experience has proved the utility of gathering them by hand, or of shaking them or brushing them from the plants into tin vessels containing a little water. They should be collected daily during the period of their visitation, and should be committed to the flames, or killed by scalding water. The late John Lowell, Esq., states, that in 1823, he discovered on a solitary apple tree, the rose bugs "in vast numbers, such as could not be described, and would not be believed if they were described, or at least none but an ocular witness could conceive of their numbers. Destruction by hand was out of the question" in this case. He put sheets under the tree, and shook them down, and burned them. Dr. Green, of Mansfield, whose investigations have thrown much light on the history of this insect, proposes protecting plants with millinet, and says that in this way only did he succeed in securing his grape vines from depredation. His remarks also show the utility of gathering them. " Eighty-six of these spoilers," says he, "were known to infest a single rose bud, and were crushed with one grasp of the hand." Suppose, as was probably the case, that one half of them were females; by this destruction, eight hundred eggs, at least, were prevented from becoming matured.

During the time of their prevalence, rose bugs are sometimes found in immense numbers on the flowers of the common white weed, or ox-eye daisy, (*Chrysanthemum leucanthemum*,) a worthless plant, which has come to us from Europe, and has been suffered to overrun our pastures, and encroach on our mowing lands. In certain cases, it may become expedient rapidly to mow down the infested white weed in dry pastures, and consume it with the sluggish rose bugs on the spot.

Our insect-eating birds undoubtedly devour many of these insects.

and deserve to be cherished and protected for their services. Rose bugs are also eaten greedily by domesticated fowls; and when they become exhausted and fall to the ground, or when they are about to lay their eggs, they are destroyed by moles, insects, and other animals, which lie in wait to seize them. Dr. Green informs us that a species of dragon fly, or devil's needle, devours them. He also says that an insect which he calls the enemy of the cut worm, probably the larva of a Carabus, or predaceous ground beetle, preys on the grubs of the common dor bug. In France, the golden ground beetle, (*Carabus auratus*,) devours the female dor or chafer at the moment when she is about to deposit her eggs. I have taken one specimen of this fine ground beetle in Massachusetts, and we have several other kinds, equally predaceous, which probably contribute to check the increase of our native Melolonthians.

THE DAHLIA.

INTRODUCTION.

Though severed from its native clime,
  Where skies are ever bright and clear,
And nature's face is all sublime,
  And beauty clothes the fragrant air,
  The Dahlia will each glory wear,
With tints as bright, and leaves as green,
As on its open plains are seen.
  And when the harvest fields are bare,
She in the sun's autumnal ray,
With blossoms decks the brow of day.

MARTIN.

ABOUT ten years before the close of the last century, this favorite flower was sent from Mexico to Spain, and a few specimens were procured, in the year of its importation to that country, from Madrid, by the then Lady Bute, but through some mismanagement the species was lost, until Lady Holland obtained seed from the same city in 1804; while in 1802, another species, (*Dahlia coccinea,*) had been brought from Mexico through France; neither the latter nor the former, (*Dahlia frustranea,*) seems, however, to have attracted much attention amongst the floricultural world; and it was not until after the peace of 1815, that it became an object of professional care, when a supply was obtained in England, from France, where its cultivation had already been carried to some extent; since which period, an indefinite number of varieties has been procured by the persevering ingenuity of the florist, and a monomania for this flower existed for many years unsurpassed in inveteracy, save by the extraordinary "Tulipomania" of the seventeenth century. This has in some degree subsided, and the Dahlia is taking its proper rank as a deservedly esteemed flower, blooming at a season of the year when the number of flowering plants in the open garden is very limited.

The name of Dahlia was given to it in honor of Dahl, a Swedish botanist and a pupil of Linnæus; there was an attempt to change it to Georgina, and on the continent this has prevailed to a considerable extent; but in England and this country, it has been entirely rejected.

REQUISITES OF A PERFECT FLOWER.

THE following characteristics are agreed upon by the London Floricultural Society as necessary to the perfection of the Dahlia:—

1st. The general form should be that of about two thirds of a sphere, or globe. The rows of petals forming this globe should describe unbroken circles, lying over each other with evenness and regularity, and gradually diminishing until they approach the top. The petals comprising each succeeding row should be spirally arranged and alternate, like the scales of the fir cone, thereby concealing the joints and making the circle more complete.

2d. The petals should be broad at the ends, perfectly free from notch or indention of any kind, firm in substance, and smooth in texture. They should be bold and free, and gently cup, but never curl or quill, nor show the under sides; they should be of uniform size, and evenly expanded in each row, being largest in the outer rows, aud gradually and proportionately diminishing until they approach the summit, when they should gently turn the reverse way, pointing towards and forming a neat and close centre.

3d. The color in itself should be dense and clear; if in an edged flower, concentrated and well defined; and in both cases penetrating through the petal with an appearance of substance and solidity.

4th. Size must be comparative.

PROPAGATION.

THE Dahlia may be propagated from tubers, by slips or cuttings, or from seed.

Propagation from Seed.

This method is now seldom practised, except by those who desire to obtain new varieties by hybridising between two distinct species or choice varieties. The proper time for sowing the seed is in March or April, in light soil in shallow boxes or pans, which are placed in a moderate hot bed to promote their germination; though some florists think that plants as vigorous, if not more so, may be obtained from seed sown in a warm and well-sheltered border toward the end of April, or in the early part of May, provided the young plants are protected during the night and guarded from casual frosts; or the seed may be sown in pans in March in the house, and put out in the open air on mild days, to accustom them to the external atmosphere. In any treatment, when the seed leaves are fully developed, they must be allowed plenty of fresh air, or placed in a cold frame, taking care that they are put as near as possible to the glass, to prevent their being drawn and growing lanky; they may also be potted singly, or three or four together, as soon as they will bear handling. When they have four leaves, they may be treated in every respect as old plants, and from the twentieth of May to the middle of June, they may be planted where it is intended they should flower.

Seed Gathering.—The seed should be collected in September from dwarf plants, where no preference exists on other accounts; and, when double varieties are principally sought for, from semi-double flowers. Seeds procured from those florets, which have changed their form, are supposed to have a greater tendency than the other to produce plants with double flowers.

Propagation by Tubers, or Slips, and by Grafting.

This is the mode most commonly adopted for the propagation of this favorite plant, and the operation is begun in March or April, by removing the tubers from the place where they have been deposited during the winter, and putting them in pots, or in loose earth on a mild hot bed. The crown of each tuber is left uncovered to permit each shoot to develop itself, under the full influence of the atmospheric air. When the shoots have attained the length of about three inches, they are cautiously separated from the tuber by laying hold of the slip with the thumb and finger near its base, and gently moving it backward and forward until it comes out of its socket. Mr. Paxton recommends that where the shoots are numerous, a part of the crown of the tuber should be invariably taken off with the shoot, a course more likely to be attended with success than by extracting the slip.

The following mode of increasing choice varieties of this favorite flower was discovered by Mr. Blake, of Kensington Gore, and is now commonly practised :—

Select a good tuber of a single sort, taking special care that it has no eyes; then, with a sharp knife, (for a dull edge would mangle the fleshy root, make it jagged, and so prevent a complete adhesion of the scion and stock,) cut off a slice from the upper part of the root, making at the bottom of the part so cut a ledge wherein to rest the graft. This is done because you cannot tongue the graft as you would do a wood shoot; and the ledge is useful in keeping the cutting fixed in its place while you tie it. Next cut the scion, (which should be strong, short jointed, having on it two or more joints or buds,) sloping to fit, and cut it so that a joint may be at the bottom of it to rest on the aforesaid ledge; a union may be effected without the ledge, provided the graft can be well fixed to the tuber, but the work will not then be so neat. It is of advantage, though not absolutely necessary, that a joint should be at the end of the scion; for the scion will occasionally put forth new roots from the lower joint; the stem is formed from the upper joint; therefore procure the cuttings with the lower joint as near together as possible.

After the graft has been tied, a piece of fine clay, su.h as is used for common grafting, must be placed round it; then pot the root in fine mould in a pot of such a size as will bury the graft half way in the mould; place the pot in a little heat in the front of a cucumber or melon frame, if you chance to have one in work at the time; the front is to be preferred, for the greater convenience of shading and watering which are required. A striking glass may be put over the graft, or not, at pleasure. In about three weeks, the root should be shifted into a large pot, if it be too soon to plant it in the border, which will probably be the case, as the plant cannot go out before April or May, so that the shifting will be very essential to promote its growth till the proper season of planting out shall arrive.

Treatment of Slips.—The shoots having been carefully separated from their parent tuber, they are immediately placed in thumb pots, filled with light soil, not inserting each more than an inch deep; when this is done, the pots are plunged in the hot bed. When they have filled these small pots with roots, they are shifted into others, which may serve them until the time for planting, unless that be protracted by unfavorable weather; in which contingency it will be desirable to remove them again into a size larger, to allow the roots to grow more freely, and to prevent their becoming a close and compact mass, which would be highly detrimental to their vigorous development, and the future health of the plant, when consigned to the open ground. Numerous shoots are emitted from the same tuber in succession, and these are treated in precisely the same manner when arrived at the proper length. They must be shaded from the sun while making roots, and protected from vapor and frost. The best compost for the Dahlia in pots is a mixture of sifted decayed hotbed dung, light virgin loam, and pure white sand, in equal quantities.

Situation and Preparation of the Soil.

The natural habitat of the Dahlia is, we are informed, in a rather light soil and on open plains. English cultivators recommend a sheltered situation; that is, sheltered from high winds, which break and shatter their lateral branches, however much they may be strengthened and supported by stakes; yet fully exposed to the sun, and where they can have the advantage of a free circulation of air, the soil naturally damp, rich, of good depth, and on a dry bottom. The soil, however, is rarely so good that it cannot be improved for the purpose for which it is desired, and it is recommended that those who would grow the Dahlia to perfection should trench the ground in November, previous to its being required, by first removing the soil to the depth of twelve inches, and replacing it with equal portions of good yellow loam and peat earth; and then trenching it again to the depth of two feet, mixing the original sub-soil an the loam and peat thoroughly together, with a large quantity of stable manure, thoroughly decayed, or it will be injurious. This may seem an expensive process, but once done it will need no further preparation for many years, except the occasional addition of manure.

N. B.—In a strong clay soil, enriched with well-decayed manure, the Dahlia produces the largest *flowers;* in a light soil, the *plant* grows to a great size, but the flowers are comparatively small.

TREATMENT.

THOSE who have no hot bed wherein to start their Dahlias into a growing state, may do so with equal success, and may obtain even more vigorous and better-blooming plants than those which are excited by artificial heat, by planting them in March or April in a box of light soil or decayed leaves, keeping it in a moist state, and exposing them to the full heat of the sun throughout the day, and taking them in-doors at night. When the shoots are three or four inches long, they may all, except one, be taken off close to the tuber, and treated as slips; but if you can divide the tuber into as many pieces as there are shoots, it is to be preferred.

Planting Out.

There are few situations, in the Middle and Northern States, where Dahlia plants can be planted out with safety before April, May, or the early part of June. When the operation is performed, the plants, if on beds by themselves, which is desirable, should be set in rows not less than six feet apart each way. Due regard must be had to the respective heights of the plants and the colors of their flowers; if on a bed where they are to be viewed from all sides, the tallest-growing kinds should be placed in the centre; if to be seen only from the front, the loftiest must be set at the back; and, in reference to colors, so arranged that they will produce a harmonious effect as a mass. Your plants, if well grown, will be from eighteen inches to twenty-four in height, when planted, and should be supported by stakes immediately; when they are full two feet high, the top of the leading shoots, or upright stem, should be cut off to induce the plant to throw out laterals.

It is a very common error to keep the Dahlia in pots too small for the quantity of roots the plant has formed, and the evil consequences of this are increased in seasons when it is most desirable they should be avoided; for if the weather be so unfavorable as to put off the period of planting out, the roots have been meanwhile increasing, and filling up the pot, so that when the plant is taken out to be set in the open ground, the ball of earth cannot be removed without breaking some of the fibres; and, fearful of doing this, many persons plant them without disturbing it, and the result generally is, that the plant does not begin to grow vigorously until near the time when it ought to be in flower. It is better, indeed, to break some of the fibres, and get away the dried and baked earth from around the roots; for though it seems to give a violent check to the growth of the plant, it will, when it has recovered, thrive far better than those planted with the ball entire; it is, however, preferable to avoid the necessity for the latter plan, or the alternative of breaking the roots, by planting them in pots of a larger size than those commonly used. The crown of the tuber should be placed at least three inches below the surface of the soil in planting out.

Mulching and Watering.

When the plants are two feet high, remove the earth from around the base of the stems to the depth of three or four inches; supplying its place with well-decomposed manure, which must be slightly covered with earth; in dry weather, the plants must be watered through this mulching twice a week at least, or every other day, according to the state of the weather; and this should be done in the evening. The

Dahlia is greatly benefited by this system of mulching and watering,; for, unlike many other kinds of plants, it seeks its nourishment chiefly from the surface of the soil; and its roots will be found, in favorable circumstances, to be clustered together near it. Throughout the summer it is also advantageous to the plants to have the earth around the roots carefully loosened by the use of a fork, from time to time

AFTER CULTURE.

Dahlias should never be pruned until the bloom buds show, and then but few branches should be cut out, and only such as are growing across others. The buds should be thinned, for it is by these that the strength of the plant gets exhausted. By removing all that are too near one to be bloomed, and all those that show imperfections enough to prevent them being useful, much strength will be gained by the future flowers. So, also, by pulling off the blooms themselves, the moment they are past perfection, instead of letting them seed.

Winds and sun are both detrimental; and the practice of fixing the blooms in the centre of a flat board, and covering them with glass or flower pots, as they may want light or shade, is becoming general. The more easy way is to use a paper shade for any particular fine bloom; for however the flowers may be coaxed and nursed under cover, a stand of blooms, grown finely, and merely shaded from the hottest sun, will beat all others in brilliancy, and in standing carriage, and keeping. It is right to go round the plants, and, wherever there is a promising bud or bloom, to take away all the leaves and shoots that threaten to touch it as they grow; take off also the adjoining buds; and if the weather be windy, make it fast to a stick or one of the stakes, that it may not be bruised or frayed; shade it from the broiling sun, and it will so profit by the air and night dews, as compared with the bloom under pots and glasses, that if the growth be equal, the blooming will be superior. Nevertheless people will cover; and where there is a disposition to a hard eye, it will hardly come out perfect unless it is covered. As the end of September approaches, or as soon as you have done with the bloom, earth up the plants, in order that when the frost comes it may not reach the crown.

Preserving the Roots.

The plants may be raised without injury, immediately after the blooms are cut off by the frost, provided that they are hung up in a dry and ordinarily protected situation, with the roots uppermost, if care is taken to leave six or seven inches of the stem attached to each tuber; this may be done without the slightest fear of their withering from having been lifted in a green state. As the winter advances, and the tubers become matured and firm, the ordinary modes of protection against frost may be resorted to.

Treatment when Flowering.

When the buds of your Dahlias begin to appear, you must take them off until you think the plants have attained their full vigor, and then permit only every third bud to grow to maturity; by doing this, it is true, you will not have so numerous a show of flowers, but those which you have, will attain the highest state of perfection your plants are capable of; taking into account their situation and previous treatment, and, what is of paramount importance, the character of the season. In the treatment of flowers grown for exhibition at flower shows, it is a common practice to bind down the disk of the flower towards the earth, by which, it is said, the flowers are rendered more perfect in form, and richer in color. When in flower, the bloom should be shaded from the sun, during the hottest parts of the day.

Striped Varieties.—The striped kinds have a tendency to "run," as it is termed, into self-colored flowers, if not carefully treated, and almost invariably do so when planted in rich soil; the best mode of keeping them "clean," that is, in their prime estate as striped flowers is to plant them in poor soil.

Autumnal and Winter Treatment.

It is the practice with many persons to take up their Dahlia roots as soon as the shrubs are cut down by the frost; this is not desirable, because if the tubers are taken up before their vital powers are in a

quiescent state, they are more easily injured by the dryness of the
atmosphere into which they are to be removed, and which it is necessary they should be able to bear without shrivelling; as in a moist
atmosphere they are apt to become mildewed and mouldy; therefore,
it is best about the end of September to cover the stems and some
distance round with earth and littery dung, about six inches thick, so
as to protect the crown of the tuber from injury by the early frosts;
and allow them to remain in the ground till November, when they
must be taken up and spread singly in a dry open shed for a few days,
not allowing the sun to shine upon them, and turned occasionally
during this period, so that they may be dried gradually; as, if dried
too quickly, they shrivel, or too slowly, they become rotten.

When sufficiently dry, clear away the earth from them, and place
them in a dry under-ground cellar, where the frost is not likely to
reach them; and these should be examined throughout the winter
from time to time, and if there be the least symptom of damp upon
the tubers, they should be carefully wiped with a dry cloth, and
receive almost daily attention. Should you not have the convenience
of such a cellar, you must store them in a pit in the garden, which
must be prepared in a dry spot, and be of sufficient capacity to hold
all your tubers. Having dug the pit, cover the bottom with dry
ashes, then pile the roots thereon, tier upon tier, so as to form a ridge;
then cover them with plenty of straw, and form a ridge of earth over
them of the thickness of twelve or fourteen inches.

FINIS